Chemical Reagents for Protein Modification

Volume I

Authors

Roger L. Lundblad, Ph.D.
Professor of Pathology
and Biochemistry
Dental Research Center
University of North Carolina
Chapel Hill, North Carolina

Claudia M. Noyes, Ph.D.
Research Associate
Department of Medicine
University of North Carolina
Chapel Hill, North Carolina

CRC Press, Inc.
Boca Raton, Florida

T P
4 5 3
. P.7
L 86
1 9 8 4
v o l. 1

Library of Congress Cataloging in Publication Data

Lundblad, Roger L.
  Chemical reagents for protein modification.

  Bibliography: p.
  Includes index.
  1. Proteins. 2. Chemical tests and reagents.
I. Noyes, Claudia M. II. Title.
TP453.P7L86   1984   661'.8   83-15076
ISBN-0-8493-5086-7 (v. 1)
ISBN-0-8493-5087-5 (v. 2)

© 1984 by CRC Press, Inc.
Second Printing, 1984
Third Printing, 1985
Fourth Printing, 1985
Fifth Printing, 1988

International Standard Book Number 0-8493-5086-7 (v. 1)
International Standard Book Number 0-8493-5087-5 (v. 2)

Library of Congress Card Number 83-15076
Printed in the United States.

PREFACE

The contents of this book are focused on the use of chemical modification to study the properties of proteins in solution. Particular emphasis has been placed on the practical laboratory aspects of this approach to the study of the relationship between structure and function in the complex class of biological heteropolymers. As a result, little emphasis is given to the individual consideration of the functional consequences of chemical modification.

The authors are indebted to the many investigators who have allowed us to include their data in this book. Such cooperation has permitted us to discuss many aspects of the specific modification of protein molecules. We are also indebted to the many journals that have allowed us to reproduce copywritten material from many laboratories.

The authors wish to express their graditude to Ms. Reneé Williams for help in the preparation of the text and to Ramona Hutton-Howe, Terri Volz, David Rainey and their colleagues in the Learning Resources Center of the School of Dentistry at the University of North Carolina at Chapel Hill for preparation of the material for figures.

One of the authors (RLL) wishes to express his particular gratitude to past and present colleagues on the fifth floor of Flexner Hall at the Rockefeller University for their contributions to this book as well to the considerable support provided for other endeavors.

The authors' research programs are supported by grants DE-02668, HL-06350 and HL-29131 from the National Institutes of Health.

Roger L. Lundblad
Claudia M. Noyes

THE AUTHORS

Roger Lauren Lundblad, Ph.D. is Professor of Pathology and Biochemistry in the School of Medicine and Professor of Oral Biology in the Department of Periodontics in the School of Dentistry at the University of North Carolina at Chapel Hill. He also serves as Associate Director for Administration of the Dental Research Center.

Dr. Lundblad received his undergraduate education at Pacific Lutheran University in Tacoma, Washington and the Ph.D. degree in biochemistry at the University of Washington in 1965. Prior to joining the faculty of the University of North Carolina in 1968, Dr. Lundblad was a Research Associate at the Rockefeller University.

Dr. Lundblad's research interests are in the use of solution chemistry techniques to study protein-protein interaction with particular emphasis on the proteolytic enzymes involved in blood coagulation and the proteins synthesized and secreted by salivary glands.

Claudia Margaret Noyes, Ph.D. is a Research Associate in the Department of Medicine at the University of North Carolina at Chapel Hill.

Dr. Noyes received her undergraduate education at the University of Vermont and the Ph.D. degree in chemistry from the University of Colorado in 1966. Prior to joining the faculty at the University of North Carolina at Chapel Hill, Dr. Noyes was a Research Associate at the University of Chicago and a Research Chemist at Armour-Dial.

Dr. Noyes's current research interests include structural studies of proteins involved in blood coagulation and high performance liquid chromatography of peptides and proteins.

TABLE OF CONTENTS

Volume I

TABLE OF CONTENTS

Volume II

Chapter 1

THE CHEMICAL MODIFICATION OF PROTEINS

The overall thrust of this book concerns the study of the solution chemistry of proteins using covalent modification of the various functional groups present in the protein. It is the purpose of this chapter to introduce the concept of the *specific* chemical modification of protein and how existence of specific chemical modification is established. Together with this subject, factors which influence the reactivity of the various functional groups in a protein will be considered.

We will define specific chemical modification as a chemical reaction which results in the *quantitative*, covalent derivatization of the functional group (i.e., primary amine, sulfydryl group, imidazole ring) of a *single, unique* amino acid residue in a protein without any demonstrable effect on either any other functional groups or the conformation of the molecule. This is an ideal pursuit which should be our foremost objective as organic chemists interested in the relationships between structure and function in a complex heteropolymer. There are several pragmatic considerations which tend to compromise our efforts in obtaining this ideal. First, few reagents are absolutely specific for a given type of functinal group (i.e., sulfhydryl, phenolic hydroxyl, etc.) let alone for a specific residue within a functional group class. The present status of analytical capabilities precludes unambiguous detection of loss less than 3 to 5% of an amino acid unless that modification is associated with the appearance of a derivative (i.e., S-carboxymethylcysteine, 3-nitrotyrosine) which can be easily identified. Secondly, it is unlikely that chemical modification of a single amino acid residue can be accomplished without *any* concomitant conformational change. Although this is a frequent criticism of chemical modification, it is, in general, *without significant merit* in the consideration of *specific chemical modification*. Admittedly, few investigators rigorously evaluate the extent of conformational change which occurs as a result of chemical modification but in such instances only minor change is seen.[1-3]

Establishment of the accomplishment of specific chemical modification is not a trivial process. There are a number of criteria to be fulfilled which are considered below.

Establishing the stoichiometry of modification is a relatively straightforward process. First, the molar quantity of modified residue is established by analysis. This could be spectrophotometric as, for example, with the trinitrophenylation of primary amino groups, the nitration of tyrosine with tetranitromethane or the alkylation of tryptophan with 2-hydroxy-5-nitrobenzyl bromide or by amino acid analysis to determine either the loss of a residue as, for example, in photo-oxidation of histidine and the oxidation of the indole ring of tryptophan with N-bromosuccinimide or the appearance of a modified residue such as with S-carboxymethylcysteine or N^1- or N^3-carboxymethylhistidine. In the situation where spectral change or radiolabel incorporation is used to establish stoichiometry, analysis must be performed to determine that there is not a reaction with another amino acid. For example, the extent of oxidation of tryptophan by N-bromosuccinimide can be determined spectrophotometrically but amino acid analysis is *required* to determine if modification has also occurred with another amino acid such as histidine.

In case of specific modification, it must be established that the modification of one residue mole per mole of protein (or functional subunit) has occurred without modification of another amino acid (e.g., modification has only occurred with lysine and not with tyrosine). This must be established with each study. The reaction pattern of a given reagent with free amino acids or amino acid derivatives does not necessarily provide the basis for reaction with such amino acid residues in protein. Furthermore, the reaction pattern of a given reagent with one protein cannot necessarily be extrapolated to all proteins. It is of critical importance to appreciate that the results of a chemical modification can be markedly affected by reaction conditions (e.g., pH, temperature, solvent and/or buffer used, degree of illumination, etc.). Establishment of stoichiometry does not necessarily mean that this modification has occurred at a unique residue (unique in terms of position in the linear peptide chain — not necessarily unique with respect to reactivity). It is, of course, useful if there is a change in biological activity (catalysis, substrate binding, ion-binding etc.) which occurs concomitant with the chemical modification (see Figure 1). Ideally, one would like to establish a direct relationship (i.e., 0.5 mol/mol of protein with 50% activity modification; 1.0 mol/mol of protein with 100% activity modification). This occurs frequently with affinity labels such as the peptide chloromethyl ketones[5,6] or with reaction at the active cysteinyl residues in sulfhydryl proteases.[7-9] More frequently, one might have the situation where there are several moles of a given amino acid modified. It may be possible to establish a relationship between the extent of chemical modification and change in biological activity during the early stages of the reaction but not during the later stages.

It has proved convenient to extend the linear relationship to the point of intersection with the axis denoting extent of modification (the x-axis in this example). This approach is widely used but only rarely do the investigators take the next step which would be to establish that there is a homogeneous population of modified protein. For example, the modification of a specific carboxyl group (Asp-101) in lysozyme[10] is associated with a change in chromatographic behavior which can be used to isolate the modified protein. The other approach to establishing that a basically homogeneous preparation of modified protein exists would be to obtain a single peptide which contains the great majority of the modified residue (see below).

The approach advanced by Ray and Koshland[11] is based on establishing a relationship between the rate of the loss of biological activity and the rate of the modification of amino acid residues. This basically involves the comparison of the second-order rate constants for the modification of ne or more amino acid residues and the loss of biological activity.

The statistical approach advanced by Tsou[12] is based on establishing a relationship between the number of residues modified and the change in biological activity. As noted by Tsou, this approach is quite valuable when it is difficult to accurately determine the rate of functional group modification as for example with N-bromosuccinimide or 2-hydroxy-5-nitrobenzyl bromide. With this approach it is necessary to determine the relatioship between the number of residues (functional groups) modified and biological activity. In Tsou's least complicated example, the biologically essential groups are all of the same type and both essential and nonessential groups are modified at the same rate. Assuming that the modification of any essential group results in the loss of activity, the fraction of biologically active protein remaining will be equal to the fraction of activity remaining (denoted as a). In the situation where there is a single essential group, the fraction of essential groups remaining after any period of modification (denoted as x_c) will be equal to a. In the situation where the number of essential groups is i (by definition greater than 1) among all functional groups of type X, the fraction of each essential group remaining after a period of modification will be x_c. Only those proteins which have all their essential groups intact will retain full activity. Therefore:

$$a = x_c^i \text{ or } a^{1/i} = x$$

When all groups of type X react at the same rate, then x_c will be equal to the fraction of the overall fraction of unmodified type X groups and

$$a^{1/i} = x$$

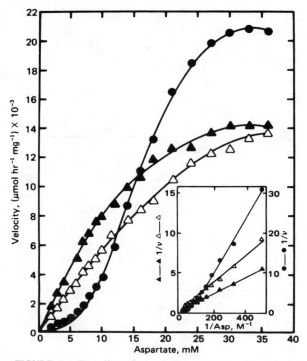

FIGURE 1. The effect of nitration on the cooperative behavior of aspartate transcarbamylase. Shown is the effect of aspartate concentration on aspartate transcarbamylase activity with the reconstituted native enzyme (●) and with the enzyme reconstituted with the nitrated catalytic subunit in the presence (△) and absence (▲) of CTP. (From Landfear, S. M., Evans, D. R., and Lipscomb, W. N., *Proc. Natl. Acad. Sci. U.S.A.*, 75, 2654, 1978. With permission.)

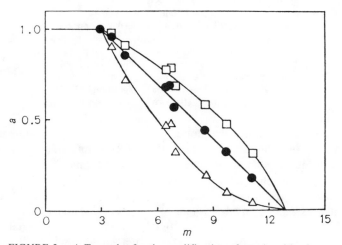

FIGURE 2. A Tsou plot for the modification of pepsin with trimethyloxonium fluoroborate. Shown is a plot of (a) vs. (m) where a is the remaining catalytic activity and m is the number of methyl groups incorporated from reaction with trimethyloxonium fluoroborate. The line for i = 2 (●) is the least-squares straight line for these points. The lines for i = 1 (△) and i = 3 (□) are the theoretical curves based upon the i = 2 line. (From Paterson, A. K. and Knowles, J. R., *Eur. J. Biochem.*, 31, 510, 1972. With permission.)

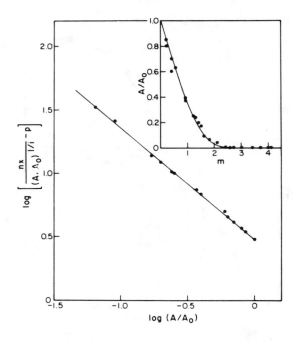

FIGURE 3.   A Tsou plot for the modification of pyridoxamine phosphate oxidase by diethylpyrocarbonate. The data
were plotted using the following equation:

$$\log\left[\frac{nx}{(A/A_0)^{1/i}} - p\right] = \log(n - p) + \left(\frac{\alpha - 1}{i}\right)\log(A/A_0)$$

where A/A_o is the fraction of enzyme molecules retaining full
activity, n is the number of modifiable residues of type X
conisting of p residues, of which i are essential, react with
the reagent at a pseudo-first-order rate constant k_1 and $n -$
p residues which are not essential reacting at a pseudo-first-
order rate constant $k_2(-\alpha k_1)$, and x is the number of residues
remaining after reaction with reagent. The data are plotted
assuming that in the above equation $n = 4$ and $p = i = 1$.
The inset describes the relationship between the number of
histidyl residues modified per mole of enzyme(m) and A/Ao.
(From Horiike, K., Tsuge, H., and McCormick, D. B., *J.
Biol. Chem.*, 254, 6638, 1979. With permission.)

The use of this approach requires the plotting of log a vs. log x; the slope of the resulting
line yields i. A number of investigators plot a(activity) vs. m(residues modified).

A recent thoughtful contribution[13] from Horiike and McCormick's laboratory has explored
the approach of relating changes in activity to extent of chemical modification. These
investigators state that the original concepts which form the basis of this approach are
sound[11,12] but that extrapolation from a plot of activity remaining vs. residues modified is
not necessarily sound. Such extrapolation is only valid if the "nonessential" residues react
much slower (rate at least 100 times slower). Given a situation where all residues within a
given group are equally reactive toward the reagent in question, the number of essential
residues obtained from such a plot is correct only when the total number of residues is equal
to the number of essential residues which is, in turn, equal to 1.0. However it is important
to emphasize that this approach is useful when there is a difference in the rate of reaction
of an *essential* residue or residues and all other residues in that class as is the example in

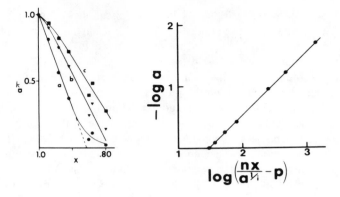

FIGURE 4. The left figure is a Tsou plot for the modification of arginine residues in transketolase by phenylglyoxal. The abscissa is a ratio of unmodified arginine residues to the total number of arginine residues. The ordinate is a where a is the fraction of activity remaining and i is a small integer. A linear segment is generated in the case where p residues including i essential residues react at a rate k and n − p residues react with rate αk. The following equation adapted from Tsou:

$$nx = pa^{1/i} + (n - p)\, a^{\alpha/i}$$

reduces to

$$a^{1/i} = \frac{nx - (n - p)}{p}$$

when $\alpha \ll 1$. This gives a straight line with the x intercept equal to n − p/n. This is represented by the extrapolation of the linear portion of the curve to the x axis: a, i = 1, b, i = 2, c, i = 3. The best fit is provided by i − 1 and in this case p = 4—5 residues/active site. The right figure describes the determination of α (the constant relating the reaction rates of rapidly and slowly reacting residues). The following equation

$$nx = pa^{1/i} + (n - p)\, a^{\alpha/i}$$

adapted from Tsou can be rearranged as:

$$\log (nx/a^{1/i} - p) = \log (n - p) + [(\alpha - 1)/i]\, \log a,$$

where i and p are determined as described above. α is determined from the slope of the resulting line. In this situation, $\alpha = 0.023$ implying that the rapidly reacting residues have a rate constant approximately 40-fold greater than the slowly reacting residues. (From Kremer, A. B., Egan, R. M., and Sable, H. Z., *J. Biol. Chem.*, 255, 2405, 1980. With permission.)

the modification of histidyl residues with diethyl pyrocarbonate in lactate dehydrogenase,[14,15] and pyridoxamine-5′-phosphate oxidase.[16] Some recent examples of the application of Tsou plots to specific chemical modification are presented in Figures 2 to 4.

Unequivocal support for *specific amino acid modification* can only be obtained from establishing that reaction occurred at a unique (in terms of position on the peptide chain) amino acid residue. This generally involves the isolation of a peptide fragment containing the modified residue. A decade ago this was a very laborious task which required substantial amounts of protein. Recent developments in the application of high-performance liquid chromatography to the analysis of peptides (reviewed in Chapter 3) have greatly decreased the amount of time and materials required for this analysis. It is of critical importance that the investigator maintain very careful records of the yield of modified residue at each fractionation step in order to be certain that the modified peptide is truly representative of the stoichiometrically modified site on the protein.

The establishment of a definite, unambiguous relationship between a specific chemical modification and a change in the catalytic activity, a change in the nature/magnitude of an allosteric response, or a change in the characteristics of a specific binding site is, in general, more difficult than one would expect considering the relatively large number of papers purporting to have elucidated such a relationship. It is difficult to completely exclude the argument that "activity" has been modified strictly secondary to the placement of steric bulk "on" a functional group as opposed to a direct effect on the participation of the functional group in the "activity". This argument was advanced to explain, in part, the effect of the modification of the active site serine of chymotrypsin with diisopropylphos-phorofluoridate. Subsequent work which showed that conversion to dehydroalanine also resulted in the complete loss of enzyme activity obviated this particular argument.[19] Koshland and Neet also described the formation of "thiol-subtilisin", a derivative of subtilisin where the active site serine has been converted to a cysteine residue.[20] Although a cysteinyl residue functions at the active site of sulfhydryl proteases such as papain,[7] ficin,[9] and streptococcal proteinase,[6] the analogous derivative of subtilisin, thiol-subtilisin, has, at best, 1% of the enzymatic activity of subtilisin. It is of interest to note that conversion of the active site cysteine to a serine residue in papain resulted in a completely inactive enzyme.[21] In both of the studies a "functionally-conservative" *chemical* mutation resulted in the essentially complete loss of catalytic activity.

One of the better approaches to the problem of relating changes in "activity" to a specific chemical modification is being able to demonstrate that the reversal of modification (see Figures 5 and 6) (or the change in the properties of a functional group which are secondary to specific modification) is directly associated with the reversal of the change(s) in biological activity. Demonstrating that the "effects" of a specific chemical modification are reversible lends strong support *against* the argument that such "effects" are a result of "nonspecific" conformation change.

In spite of the various arguments against the merit of such studies, a careful study of a specific chemical modification can provide extensive information about the organic chemistry of a class of heteropolymers (protein). The process of the elucidation of the differences in reactivities of various residues in a protein can provide considerable insight into the possible function of such residues. The differences in the rate of modification of the active site histidine in lactate dehydrogenase and α-N-acetylhistidine[14] provide some insight into the increase in nucleophilicity associated with the presence of this residue at the active site of an enzyme. Our laboratory[24] has also been able to use reaction with diethylpyrocarbonate as a measure of the change in the nucleophility of the active site histidine associated with the conversion of bovine α-thrombin to β-thrombin. Both of the above studies can also be considered as investigations into the class(es) of functional groups critical for catalysis by a particular enzyme.

In the study of the effect of chemical modification on enzymes, it is of critical importance to differentiate between modifications which have an actual effect on catalysis (the *actual process* of bond-breaking and bond-making), modifications which have an effect on the microenvironment of enzymes and modifications which have a primary effect on the process of substrate recognition (the modification of the aspartic acid residue at S_1, see Figure 7, in trypsin[25] is an excellent example of such a modification).

Another major use of chemical modification has been in the determination of the primary

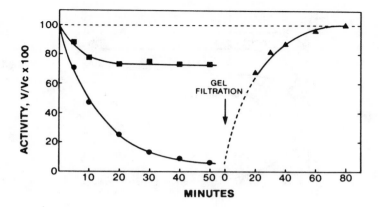

FIGURE 5. The reversible modification of pyridoxamine-5'-phosphate oxidase by 2,3-butanedione. The enzyme (2.1 μ*M*) was incubated with 10 m*M* 2,3-butanedione in the presence of 5 μ*M* flavin mononucleotide (FMN) in either 50 m*M* potassium borate, pH 8.0 (●) or 50 m*M* potassium phosphate, pH 8.0 (■). The reaction mixture in borate was passed over a G-25 Sephadex column and assayed for enzyme activity (▲). The arrow indicates the time at which the reaction mixture in borate was applied to the gel filtration column. (From Choi, J.-D. and McCormick, D. B., *Biochemistry*, 20, 5722, 1981. With permission.)

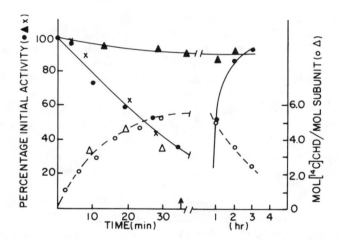

FIGURE 6. The reversible modification of ADP-glucose synthetase by 1,2-cyclohexanedione (CHD). The enzyme was incubated with 10 m*M* [^{14}C] CHD in the presence (●,○) or absence (x, △) of 50 m*M* sodium tetraborate. The control (▲) was incubated in the absence of CHD. Portions were removed at the indicated times for the determination of incorporated radioactivity (open symbols) or fructose diphosphate-stimulated enzyme activity (closed symbols). The arrow indicated the time of addition of 0.2 *M* hydroxylamine. (From Carlson, C. A. and Preiss, J., *Biochemistry*, 21, 1929, 1982. With permission.)

structure of proteins. This includes reagents such as cyanogen bromide for the chemical cleavage of specific peptide bonds, citraconic anhydride for the reversible blocking of lysine residues to restrict tryptic cleavage to arginine residues, and the reversible blocking of arginine residues with 1,2-cyclohexanedione to restrict tryptic cleavage to lysine residues. While the direct determination of amino acid sequence from intact proteins continues to be

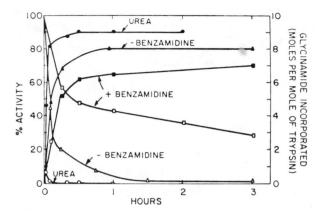

FIGURE 7.   The modification of β-trypsin with N-(3-dimethyl amino propyl)N-ethyl carbodiimide HCl (EDC) and glycinamide in the presence and absence of benzamidine and in the presence of urea as indicated in the figure. Modification of carboxyl groups was determined by the incorporation of radiolabeled glycinamide (■, ▲, ●) and catalytic activity (□, △, ○) was determined with benzoyl-arginine ethyl ester. (From Eyl, A. W., Jr. and Inagami, T., *J. Biol. Chem.*, 246, 738, 1971. With permission.)

of sufficient interest to merit publication, it is abundantly clear that the elucidation of protein structure in the future will come primarily from the sequence analysis of the cDNA of that specific protein.[26-28] This is not to say that the primary structure analysis of proteins is without value, but it is our feeling that primary value of such analysis is to obtain information regarding the site of specific chemical modification (discussed above), the site of post-translational modification (such as the formation of γ-carboxyglutamic acid, esterification or N-methylation), the site of mutation in a protein, or to obtain sufficient primary structure information for the synthesis of a suitable cDNA probe which can be subsequently used for the isolation of specific messenger RNA.

While there is considerable interest in the above areas of activity, it is our impression that there is considerably greater interest in (1) the development and use of affinity labels (see Figures 8 to 13[29-31]) to identify active site residues (this is extensively examined in Volume II, Chapter 6); (2) the development and use of structural "probes" where change in spectral (ultraviolet spectroscopy, fluorescence) or other properties (electron spin resonance — ESR with "spin-labeled" derivatives; nuclear magnetic resonance — NMR with certain compounds) can be used to assess changes in the microenvironment surrounding a specific region of the protein molecule (e.g., binding sites, active sites); and (3) the determination of secondary, tertiary, or quarternary structures of proteins (e.g., use of cross-linking agents — see Volume II, Chapter 4 and the use of functional group reactivity to differentiate between "exposed" and "buried" residues). Two of these areas are covered extensively in other chapters while some applications of structural probes are discussed below.

The use of specific chemical modification to study changes in environment has been studied over the past 30 years. There is some consensus that the study of Kirtley and Koshland[32] provided the basis for the concept of using "reporter" groups to study changes in the microenvironment surrounding a site of modification. This study used 2-bromoacetamido-4-nitrophenol to modify a limited number of sulfhydryl groups in glyceraldehyde-3-phosphate dehydrogenase. The modified protein has a λ_{max} at 390 nm ($\epsilon = 7100\ M^{-1}$ cm^{-1}) between pH 7.0 and pH 7.6. The addition of the coenzyme, NAD, caused a marked

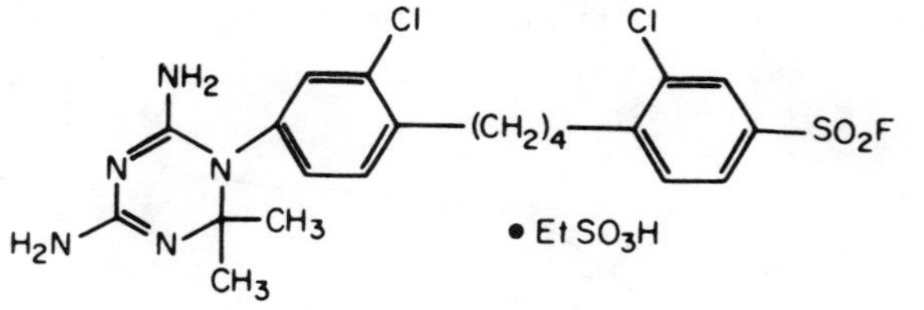

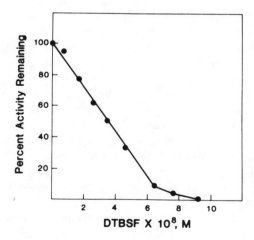

NSC 127755 (DTBSF)

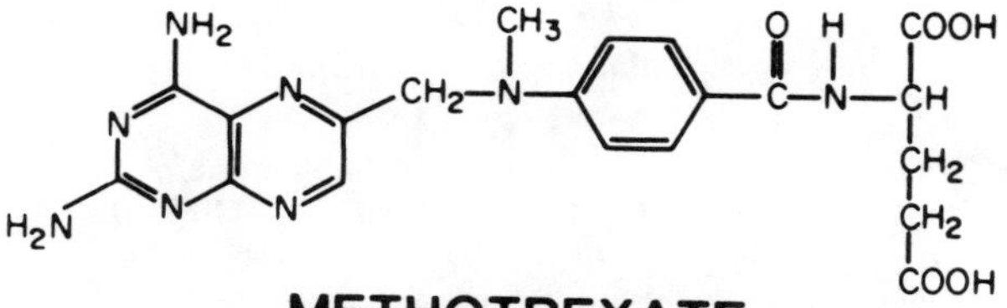

METHOTREXATE

FIGURE 8. The rationale for the development of an affinity label. The structures of 3-chloro-4[4-[2-chloro-4-(4,6-diamino-2,2-dimethyl-*S*-triazin-1-(2H,yl)phenyl]-butyl]-benzene-sulfonyl-) fluoride (DTBSF) and methotrexate. (From Kumar, A. A., Mangum, J. H., Blankenship, D. T., and Freisheim, J. H., *J. Biol. Chem.*, 256, 8970, 1981. With permission.)

FIGURE 9. The inactivation of dihydrofolate reductase by DTBSF. Dihydrofolate reductase (6.6 μM) was incubated with DTBSF at the indicated concentration in 0.1 M Tris, pH 7.2 for 2 min and then assayed for catalytic activity. (From Kumar, A. A., Mangum, J. H., Blankenship, D. T., and Freisheim, J. H., *J. Biol. Chem.*, 256, 8970, 1981. With permission.)

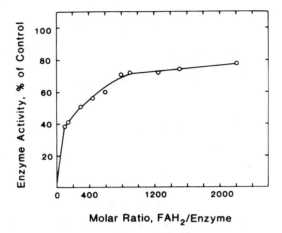

FIGURE 10.   The effect of substrate (dihydrofolate) on the
inactivation of dihydrofolate reductase by DTBSF. Dihydrofolate reductase (0.12 μM) was preincubated for 2
min with dihydrofolate at the indicated concentration for 2
min and subsequently reacted with DTBSF for 2 min and
the remaining catalytic activity determined. (From Kumar,
A. A., Mangum, J. H., Blankenship, D. T., and Freisheim,
J. H., *J. Biol. Chem.*, 256, 8970, 1981. With permission.)

change in the spectral properties (decrease in absorbance at approximately 375 nm and
increase in absorbance at approximately 420 nm) of the modified enzyme which is consistent
with a change in the microenvironment around the modified residue (increase in polarity of
medium which results in increased formation of the nitrophenolate ion). The reaction of 2-
hydroxy-5-nitrobenzyl bromide with tryptophanyl residues to yield the 2-hydroxy-5-nitro-
benzyl derivative[33] (see Volume II, Chapter 2) and the reaction of tetranitromethane with
tyrosyl residues[34] to form these 3-nitrotyrosyl derivatives (see Volume II, Chapter 3) were
developed shortly after this study and have been extensively used to study microenvironmental changes in the modified proteins (Figures 14 to 16).

The use of spin-labeled reagents to determine the conformational environment at enzyme-
active sites and other binding sites on proteins has been of considerable interest during the
past decade. Berliner and co-workers have been particularly active in this area.[35] One early
study used spin-labeled derivatives of diisopropylphosphorofluoridate to study the active site
environment of trypsin[36] (Figures 17 to 19). Subsequent studies used various spin-label
derivatives (piperidinyl nitroxide, pyrrolidinyl nitroxide and pyrrolinyl nitroxide substituent
groups) of phenylmethylsulfonyl fluoride to compare microenvironments surrounding the
active sites in α-chymotrypsin and trypsin.[37,38] These reagents have been more recently used
to study the active site of thrombin.[39,40] The preparation of spin-labeled pepsinogen has been
reported.[41] This study used a N-hydroxy succinimide ester derivative, 3-[[(2,5-dioxo-1-
pyrrolidiny)oxy] carbonyl]-2,5,-dihydro-2,2,5,5,-tetramethyl-1*H*-pyrrolyl-1-oxy, to modify
lysyl residues in pepsinogen. Coupling was accomplished at pH 7.0 (0.1 *M* sodium phosphate)
for 7 hr at 22°C resulting in the derivatization of approximately three amino groups. Figure
20 shows the change in the ESR spectrum of the modified protein upon activation (it is
noted that the modified protein retained full potential peptic activity). Figure 21 shows this
change in greater detail at pH 2.38 while Figure 22 shows an apparent lag phase when
activation is performed at pH 2.77. It is clear that the change in the "signal" generated by
the label is a measure of the rate of proenzyme to enzyme transformation as shown in Figure
23.

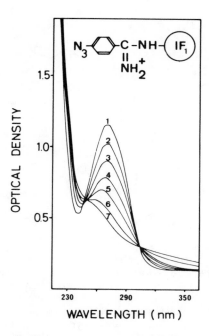

FIGURE 11. The spectra of methyl-4-azido-
benzoimidate beef heart mitochondrial coupling factor (MABI-IF$_1$). A solution of MABI-
IF$_1$ in 10 mM ammonium sulfate, pH 8.0,
was irradiated in a 1.0 cm quartz cuvette with
mineral light UVS 11. The figure shows that
the absorbance peak at 273 nm was decreased
upon irradiation (traces 1—7). (From Klein,
G., Satre, M., Dianoux, A.-C., and Vignais,
P. V., *Biochemistry,* 20, 1339, 1981. With
permission.)

Fluorescent probes have been particularly useful in the study of protein structure. A
rigorous review of this area is beyond the scope of the present work and only a limited
number of examples will be considered. One exceptional study[42] in this area used a dan-
sylpeptide chloromethyl ketone to label the active site of blood coagulation factor Xa and
then subsequently used this derivative to study the stoichiometry of interaction with the other
components of the prothrombin-activation enzyme complex (Figures 24 to 26). One of the
major problems in the labeling of proteins to obtain fluorescent derivatives is the difficulty
in achieving selective modification. One clever approach to this problem has been advanced
by Jackson and co-workers.[43] These investigators used the observation that 3-aminotyrosyl
residues (obtained by reaction with tetranitromethane followed by a reduction with sodium
hydrosulfite) are more reactive than other primary amino functions in a protein to obtain a
ation of an apolipoprotein specifically modified (Figure 27) with 5-dimethylamino-
naphthalene-1-sulfonyl chloride (dansyl chloride). These investigators were subsequently
able to use the increase in the fluorescence associated with the binding of the dansyl-
apolipoprotein to a phospholipid vesicle (Figure 28) to examine kinetics of association of
phospholipid with apolipoprotein (Figure 29).

To gain a thorough appreciation of the events involved in the specific chemical modification
of proteins, it is necessary to understand some of the basic organic chemistry of the functional
groups on the protein and the reagents which react with these functional groups.

Most of the reactions which will be described in the following pages are not actually a

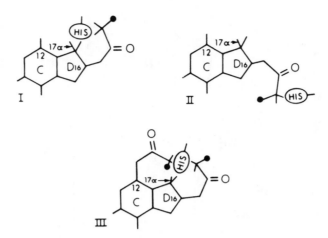

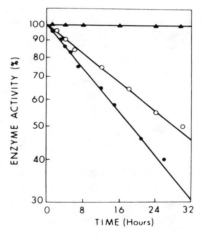

FIGURE 12.   The design of affinity labels for the modification of estradiol-17β-dehydrogenase. The two bromoacetoxy groups of 12β-(bromoacetoxy)-4-estrene-3,17-dione and 16α-(bromacetoxy)estradiol 3-methyl ether would overlap the catalytic site of estradiol 17β-dehydrogenase and would both "affinity" label a single histidine residue which proximates the steroid 17α-hydrogen. For clarity, the 18-methyl group has been omitted. (From Chin, C.-C., Murdock, G. L., and Warren, J. C., *Biochemistry,* 21, 3322, 1982. With permission.)

FIGURE 13.   The inactivation of estradiol 17β-dehydrogenase by 100μM 12 β-(bromoacetoxy)-4-estrene-3,17-dione (●), 100 μM 12β-(bromoacetoxy)-4-estrene-3,17-dione and 100 μM estradiol(○), and either a control with enzyme alone or with 100 μM bromoacetate (▲). (From Chin, C.-C., Murdock, G. L., and Warren, J. C., *Biochemistry,* 21, 3322, 1982. With permission.)

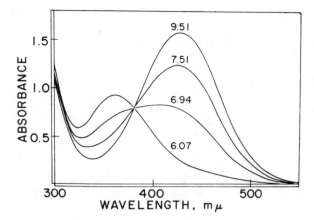

FIGURE 14.   The absorption spectrum of *N*-acetyl-3-nitrotyrosine (0.25 m*M*) in 0.2 *M* Tris, 0.2 *M* acetate, 0.5 *M* NaCl at the pH indicated in the figure. (From Riordan, J. R., Sokolovsky, M., and Vallee, B. L., *Biochemistry*, 6, 358, 1967. With permission.)

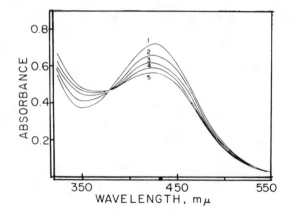

FIGURE 15.   Changes in the absorption spectrum of ni-trated carboxypeptidase in the presence of a competitive inhibitor, β-phenylpropionate. The spectra were determined in 0.2 *M* Tris, 0.2 *M* acetate, 0.5. NaCl, pH 8.0. The concentrations of β-phenylpropionate were 1 (none), 2(0.01 *M*), 3(0.025 *M*), 4(0.05 *M*), and 5(0.1 *M*). (From Riordan, J. R., Sokolovsky, M., and Vallee, B. L., *Biochemistry*, 6, 358, 1967. With permission.)

result of the reagent in question attacking the functional groups on the protein with the subsequent formation of a covalent bond but rather nucleophilic attack by the functional group on the protein on an electron-poor center such as the carbonyl carbon of an α-haloacid (e.g., iodoacetic acid).

Before going further it might be useful to define some terms that we will be using. When most of us think of acids we tend to think of substances such as hydrochloric acid, sulfuric acid, and acetic acid, substances which can donate protons in aqueous solution to form hydronium ions (H_3O^+). Likewise a base is most usually considered to be a substance (e.g., hydroxide ion — OH^-) which can accept protons. In other words, by this definition a base possesses an unshared electron pair with which it can attract and hold a proton. This is the

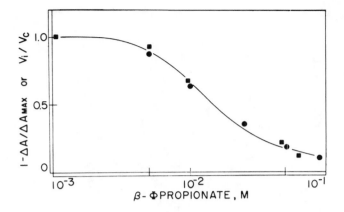

FIGURE 16.   The effect of a competitive inhibitor, β-phenylpropionate on the esterase activity (■) and the absorbance at 428 nm (●) of nitrocarboxypeptidase. $\triangle A/\triangle A_{max}$ represents the fractional decrease in absorbance at 428 nm resulting from the indicated concentration of β-phenylpropionate. $\triangle A_{max}$ is the maximal decrease in absorbance at 428 nm obtained by extrapolation to infinite concentration of β-phenylpropionate. V/Vc is the ratio of catalytic activity in the presence (V) and absence (Vc) of β-phenylpropionate. (From Riordan, J. R., Sokolovsky, M., and Vallee, B. L., *Biochemistry*, 6, 358, 1967. With permission.)

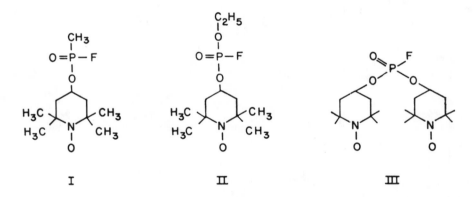

FIGURE 17.   The structures of spin-labeled derivatives used to study the conformation of trypsin as shown in Figures 19 and 20. These are various derivatives of diisopropylphosphorofluoridate. Compound I is 1-oxyl-2,2,6,6-tetramethyl-4-piperidinylmethylphosporofluoridate. (From Berliner, L. J. and Wong, S. S., *J. Biol. Chem.*, 248, 1118, 1973. With permission.)

classical Bronsted definition of acids and bases. Organic chemists find it more useful to use the Lewis definition of acids and bases. Using this definition, an acid is a substance that can form a covalent bond by accepting an electron pair and a base has an unshared electron pair. Taking this a step further then, Lewis acids are electrophilic while Lewis bases are nucleophilic.

As we have indicated above, the majority of the chemical reactions to be described later involve nucleophilic addition by a functional group on the protein molecule. Nucleophilic addition proceeds by either an S_N1 mechanism or an S_N2 mechanism.

Now we can go somewhat further and equate a nucleophilic species with an electron-rich center and an electrophilic species with an electron-deficient center. A list of the nucleophilic functional groups in proteins is given in Table 1. In general the most potent nucleophile in

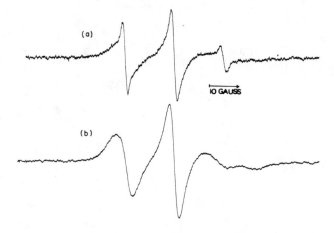

FIGURE 18. The electron spin resonance (ESR) spectra of trypsin labeled with compound I (see Figure 17). Spectrum a was obtained by modification at pH 5.5 for 10—20 hr followed by gel filtration at pH 3.5. Spectrum b was obtained by labeling for 1—2 hours at pH 7.7 followed by gel filtration, collodion bag concentration and dialysis at pH 3.5. (From Berliner, L. J. and Wong, S. S., *J. Biol. Chem.*, 248, 1118, 1973. With permission.)

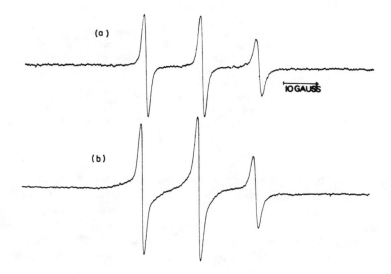

FIGURE 19. The ESR spectra of trypsin. Sample a was inert trypsin obtained from chromatographic fractionation of spin-labeled enzyme described in Figure 18. Sample b was obtained with α-β-trypsin labeled with compound I (see Figure 17) and then treated with added commercial trypsin. (From Berliner, L. J. and Wong, S. S., *J. Biol. Chem.*, 248, 1118, 1973. With permission.)

a protein is sulfur with nitrogen considerably less potent followed in potency by oxygen and carbon.

It is useful to consider some factors which influence the reactivity of nucleophilic centers in proteins. From a consideration of the three-dimensional structure of proteins the majority of polar amino acids (i.e., Lys, Arg, Gly, Asp) are located on the exterior surface of the molecule while the majority of the hydrophobic (nonpolar) residues are located in the interior

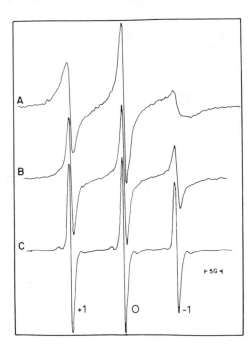

FIGURE 20.   The ESR spectrum of porcine pepsinogen and pepsin preparations. The protein derivatives have been labeled with 3-[[(2,5-dioxo-1-pyrrol-
idinyl)oxy]carbonyl]-2,5-dihydro-2,2,5,5-tetramethyl-
1H-pyrrolyl-1-oxy. Spectrum A shows spin-labeled
pepsinogen in 0.02 M sodium phosphate, pH 7,0;
spectrum B with the same solution as A with a 2.0:1
molar ratio of pepstatin brought to pH 1.5 with HCl;
spectrum C shows the spin-labeled pepsinogen solution taken to pH 1.3 for 300 min with HCl. (From
Twining, S. S., Sealy, R. C., and Glick, D. M.,
Biochemistry, 20, 1263, 1981. With permission.)

of these molecules. Thus, a gradient of polarity (dielectric constant) will exist going from
the surface of the protein into the interior. Such a gradient could also be considered to exist
in "pocket-like" indentations on the protein surface. For example, the area immediately
adjacent to the active site and substrate site S_1 in thrombin is definitely hydrophobic with
respect to the surrounding environment and the aqueous solution. This is best demonstrated
by the increase in the fluorescence of $N\alpha$-dansyl-L-arginine-N-3-ethyl-1,5-pentanediyl) amide
upon binding to the active-site region.[44]

It should follow from the above discussion that the surface of a globular protein is definitely
not homogeneous with respect to electrical charge or, more critically for our consideration,
with respect to dielectric constant. As a result of this lack of homogeneity, a variety of
surface polarities will surround the various functional groups. The physical and chemical
properties of any given functional group will be strongly influenced by the nature (e.g.,
polarity) of the local microenvironment. Changes in the polarity of the microenvironment
can have a profound effect on the dissociation of acids. For example consider the effect of
the addition of an organic solvent, ethyl alcohol, on the pKa of acetic acid. In 100% H_2O,
acetic acid has a pKa of 4.70. The addition of 80% ethyl alcohol results in an increase of
the pKa to 6.9. In 100% ethyl alcohol the pKa of acetic acid is 10.3. This is particularly
important in considering the reactivity of nucleophilic groups such as amino groups, cysteine,

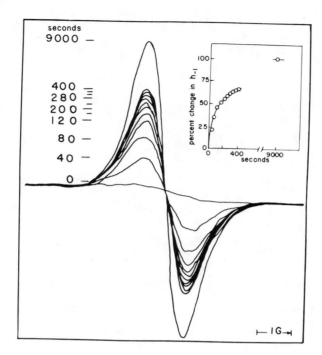

FIGURE 21.   Repeated scans of the downfield peak following activation of the spin-labeled pepsinogen preparation described in Figure 20. The ESR spectrum of the peak was scanned at 40-sec intervals following the acidification of the spin-labeled pepsinogen preparation to pH 2.38 with HCl. The insert shows the peak height vs. time. (From Twining, S. S., Sealy, R. C., and Glick, D. M., *Biochemistry*, 20, 1267, 1981. With permission.)

carboxyl groups, and the phenolic hydroxyl group. In the case of the primary amines present in protein, these functional groups are not reactive until the free base is available. In other words the proton present at neutral pH must be removed from the ϵ-amino group of lysine before this functional group can function as an effective nucleophile. A listing of the "average" pKa values for the various functional groups present in protein is also given in Table 1.

Other factors which can influence the pKa of a functional group in a protein include hydrogen binding with an adjacent functional group, the direct electrostatic effect of the presence of a charged group in the immediate vicinity of a potential nucleophile and finally, direct steric effects on the availability of a given functional group.

There is another consideration which can be in a sense considered either a cause or consequence of microenvironmental polarity. This has to do with the functional groups/ environment immediately around the nucleophilic species in question. These are the "factors" that can cause a "selective" increase (or decrease) in reagent concentration in the vicinity of a potentially reactive species. The most clearly understood example of this is the process of affinity labeling. Another situation can be related to the differences in polarity of the microenvironment around a nucleophilic center. This is considered in Volume II, Chapter 3 and is discussed by Myers and Glazer.[45] There is also the consideration that a charged reagent can be either attracted to or repelled from the vicinity of a nucleophilic center. This is easily demonstrated by the differences in the comparative rates of modification of the active site cysteinyl residue by chloroacetic acid and chloroacetamide in papain[46] and streptococcal proteinase.[8]

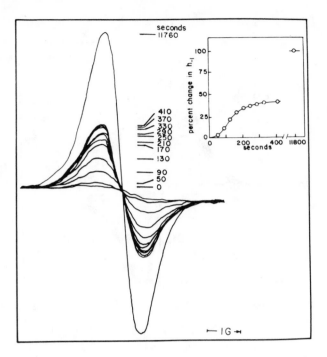

FIGURE 22.   Conformational changes in pepsinogen occurring during the activation to pepsin. This figure shows repeated ESR spectra of the downfield peak following the activation of spin-labeled pepsinogen following acidification of a 32.3 μM solution to pH 2.77. The insert shows the peak height vs. time. (From Twining, S. S., Sealy, R. C., and Glick, D. M., *Biochemistry*, 20, 1267, 1981. With permission.)

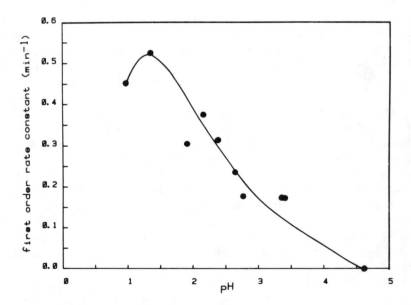

FIGURE 23.   The study of the rate of ESR changes of spin-labeled pepsinogen as a function of pH. The acidified pepsinogen solutions were analyzed as described in Figure 21. The first-order rate constants were determined from the initial 20% of the total change in spectra. (From Twining, S. S., Sealy, R. C., and Glick, D. M., *Biochemistry*, 20, 1267, 1981. With permission.)

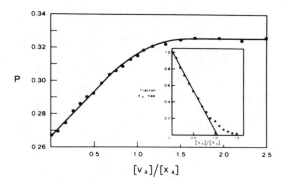

FIGURE 24. The interaction of fluorescent factor Xa with factor Va in the presence of phospholipid and calcium ions. The factor Xa had been modified with Dansyl-Glu-Gly-Arg-chloromethyl ketone. The value for fluorescence polarization (P) is plotted vs. the concentration of factor Va added. The inset shows a plot of the factor Xa free (not in complex formation as assessed by fluorescence polarization data) vs. the total factor Xa added. Extrapolation of the data indicated in the insert suggests a binding stoichiometry of 1:1 between factor Xa and factor Va in the presence of calcium ions and phospholipid. (From Nesheim, M. E., Kettner, C., Shaw, E., and Mann, K. G., *J. Biol. Chem.*, 256, 6537, 1981. With permission.)

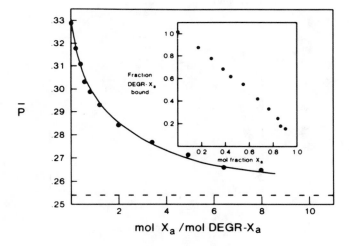

FIGURE 25. The displacement of fluorescent factor Xa (see Figure 24) from complex with factor Va in the presence of calcium ions and phospholipid by unlabeled factor Xa. Fluorescence polarization values (ordinate) are plotted vs. the ratio of unlabeled factor Xa to fluorescent factor Xa (abscissa). The inset shows the fraction of fluorescent factor Xa bound to factor Va, calculated from polarization values, vs. the mole fraction of unlabeled factor Xa. (From Nesheim, M. E., Kettner, C., Shaw, E., and Mann, K. G., *J. Biol. Chem.*, 256, 6537, 1981. With permission.)

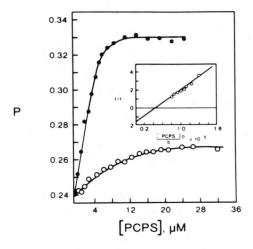

$$[\text{PCPS}], \mu M$$

FIGURE 26.   The binding of fluorescent factor Xa (see Figure 24) to phospholipid in the presence (closed circles) and absence (open circles) of factor Va. The degree of binding was assessed by the degree of fluorescence polarization (P). The inset shows a double reciprocal analysis of the binding isotherm for the factor Va-independent interaction of factor Xa with phospholipid. (From Nesheim, M. E., Kettner, C., Shaw, E., and Mann, K. G., *J. Biol. Chem.*, 256, 6537, 1981. With permission.)

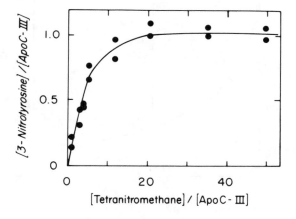

FIGURE 27.   The modification of apolipoprotein C-III (apoC-III) with tetranitromethane. The extent of nitrotyrosine formation is plotted vs. the molar excess of tetranitromethane. (From Cardin, A. D., Jackson, R. L., and Johnson, J. D., *J. Biol. Chem.*, 257, 4987, 1982. With permission.)

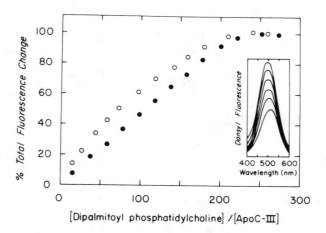

FIGURE 28. The increase in fluorescence of labeled apoC-III (the nitrated apoC-III was reduced with dithionite and subsequently reacted with dansyl chloride to provide the fluorescent derivative) (open circles) and native apoC-III (closed circles) occurring on interaction with dipalmitoyl phosphatidyl choline unilamellar vesicles. Intrinsic fluorescence was utilized with the native apolipoprotein. The inset shows the increase and wavelength shift in dansyl fluorescence emission with dansyl-apoC-III after successive additions of dipalmitoyl phosphatidyl choline. (From Cardin, A. D., Jackson, R. L., and Johnson, J. D., *J. Biol. Chem.*, 257, 4987, 1982. With permission.)

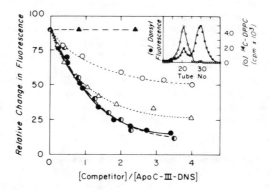

FIGURE 29. The decrease in dansyl fluorescence intensity of labeled apoC-III (see Figure 28) occurring on the addition of bovine serum albumin (▲), apolipoprotein C-I (○), apolipoprotein C-II (△), apolipoprotein A-I(●), or apolipoprotein C-III (◖). (From Cardin, A. D., Jackson, R. L., and Johnson, J. D., *J. Biol. Chem.*, 257, 4987, 1982. With permission.)

Table 1
DISSOCIATION CONSTANTS FOR NUCLEOPHILES IN PROTEINS

Potential nucleophile	pKa
Carboxyl	4.6
Imidazole	7.0
Sulfhydryl	7.0
Alpha-amino	7.8
Phenolic hydroxyl	9.6
Epsilon-amino	10.2

REFERENCES

1. **Plapp, B. V., Zeppezauer, E., and Bränden, C. I.,** Crystallization of liver alcohol dehydrogenase activated by the modification of amino groups, *J. Mol. Biol.,* 119, 451, 1978.
2. **Zisapel, N., Mallul, Y., and Sokolovsky, M.,** Tyrosyl interactions at the active site of carboxypeptidase B, *Int. J. Peptide Protein Res.,* 19, 480, 1982.
3. **Kirschner, M. W. and Schachman, H. K.,** Conformational studies on the nitrated catalytic subunit of aspartate transcarbamylase, *Biochemistry,* 12, 2987, 1973.
4. **Landfear, S. M., Evans, D. R., and Lipscomb, W. N.,** Elimination of cooperativity in aspartate trans-carbamylase by nitration of a single tyrosine residue, *Proc. Natl. Acad. Sci. U.S.A.,* 75, 2654, 1978.
5. **Shaw, E., Mares-Guia, M., and Cohen, W.,** Evidence for an active-center histidine in trypsin through use of a specific reagent, 1-chloro-3-tosylamido-7-amino-2-heptanone, the chloromethyl ketone derived from N^α tosyl-L-lysine, *Biochemistry,* 4, 2219, 1965.
6. **Schoellmann, G. and Shaw, E.,** Direct evidence for the presence of histidine in the active center of chymotrypsin, *Biochemistry,* 2, 252, 1963.
7. **Glazer, A. N. and Smith, E. L.,** Papain and other plant sulfhydryl proteolytic enzymes, in *The Enzymes,* Vol. 3, Boyer, P. D., Ed., Academic Press, New York, 1971, 502.
8. **Gerwin, B. I.,** Properties of the single sulfhydryl group of streptococcal proteinase. A comparison of the rates of alkylation by chloroacetic acid and chloroacetamide, *J. Biol. Chem.,* 242, 451, 1967.
9. **Englund, P. T., King, T. P., Craig, L. C., and Walti, A.,** Studies on ficin. I. Its isolation and characterization, *Biochemistry,* 7, 163, 1968.
10. **Yamada, A., Imoto, T., Fujita, K., Okasaki, K., and Motomura, M.,** Selective modification of aspartic acid-101 in lysozyme by carbodiimide reaction, *Biochemistry,* 20, 4836, 1981.
11. **Ray, W. J., Jr. and Koshland, D. E., Jr.,** A method for characterizing the type and numbers of groups involved in enzyme action, *J. Biol. Chem.,* 236, 1973, 1961.
12. **Tsou, C.-L.,** Relation between modification of functional groups of proteins and their biological activity. I. A graphical method for the determination of the number and type of essential groups, *Sci. Sinica,* 11, 1535, 1962.
13. **Horiike, K. and McCormick, D. B.,** Correlations between biological activity and the number of functional groups chemically modified, *J. Theoret. Biol.,* 79, 403, 1979.
14. **Holbrook, J. J. and Ingram, V. A.,** Ionic properties of an essential histidine residue in pig heart lactate dehydrogenase, *Biochem. J.,* 131, 729, 1973.
15. **Bloxham, D. P.,** The chemical reactivity of the histidine-195 residue in lactate dehydrogenase thiomethylated at the cysteine-165 residue, *Biochem. J.,* 193, 93, 1981.
16. **Horiike, K., Tsuge, H., and McCormick, D. B.,** Evidence for an essential histidyl residue at the active site of pyridoxamine (pyridoxine)-5'-phosphate oxidase from rabbit liver, *J. Biol. Chem.,* 254, 6638, 1979.
17. **Paterson, A. K. and Knowles, J. R.,** The number of catalytically essential carboxyl groups in pepsin. Modification of the enzyme by trimethyloxonium fluoroborate, *Eur. J. Biochem.,* 31, 510, 1972.
18. **Kremer, A. B., Egan, R. M., and Sable, H. Z.,** The active site of transketolase. Two arginine residues are essential for activity, *J. Biol. Chem.,* 255, 2405, 1980.

19. **Strumeyer, D. H., White, W. N., and Koshland, D. E., Jr.,** Role of serine in chymotrypsin action. Conversion of the active serine to dehydroalanine, *Proc. Natl. Acad. Sci. U.S.A.,* 50, 931, 1963.
20. **Neet, K. E. and Koshland, D. E., Jr.,** The conversion of serine at the active site of subtilisin to cysteine: a "chemical mutation", *Proc. Natl. Acad. Sci. U.S.A.,* 56, 1606, 1966.
21. **Clark, P. I. and Lowe, G.,** Conversion of the active-site cysteine residue into a dehydroserine, a serine and a glycine residue, *Eur. J. Biochem.,* 84, 293, 1978.
22. **Choi, J.-D. and McCormick, D. B.,** Roles of arginyl residues in pyridoxamine-5'-phosphate oxidase from rabbit liver, *Biochemistry,* 20, 5722, 1981.
23. **Carlson, C. A. and Preiss, J.,** Involvement of arginine residues in the allosteric activation of *Escherichia coli* ADP-glucose synthetase, *Biochemistry,* 21, 1929, 1982.
24. **Lundblad, R. L., Nesheim, M. E., Straight, D. L., Bowie, J., Mann, K. G., and Roberts, J. D.,** Bovine α- and β-thrombin. Changes in reactivity of the active site are responsible for reduced fibrinogen clotting activity of β-thrombin, manuscript submitted for publication, 1982.
25. **Eyl, A. W., Jr. and Inagami, T.,** Identification of essential carboxyl groups in the specific binding site of bovine trypsin by chemical modification, *J. Biol. Chem.,* 246, 738, 1971.
26. **Law, S. W. and Dugaiczyk, A.,** Homology between the primary structure of α-fetoprotein, deduced from a complete cDNA sequence, and serum albumin, *Nature (London),* 291, 201, 1981.
27. **Kehry, M. R., Fuhrman, J. S., Schilling, J. W., Rogers, J., Sibley, C. H., and Hood, L. E.,** Complete amino acid sequence of a mouse μ chain: homology among heavy chain constant region domains, *Biochemistry,* 21, 5415, 1982.
28. **Kurachi, K. and Davie, E. W.,** Isolation and characterization of a cDNA coding for human factor IX, *Proc. Natl. Acad. Sci. U.S.A.,* 79, 6461, 1982.
29. **Kumar, A. A., Mangum, J. H., Blankenship, D. T., and Freisheim, J. H.,** Affinity labeling of chicken liver dihydrofolate reductase by a substituted 4,6-diaminodihydrotriazine bearing a terminal sulfonyl fluoride, *J. Biol. Chem.,* 256, 8970, 1981.
30. **Klein, G., Satre, M., Dianoux, A.-C., and Vignais, P. V.,** Photoaffinity labeling of mitochondrial adenosine triphosphatase by an azido derivative of the natural adenosine triphosphatase inhibitor, *Biochemistry,* 20, 1339, 1981.
31. **Chin, C.-C., Murdock, G. L., and Warren, J. C.,** Identification of two histidyl residues in the active site of human placental estradiol 17-dehydrogenase, *Biochemistry,* 21, 3322, 1982.
32. **Kirtley, M. E. and Koshland, D. E., Jr.,** The introduction of a "reporter" group at the active site of glyceraldehyde-3-phosphate dehydrogenase, *Biochem. Biophys. Res. Commun.,* 23, 810, 1966.
33. **Loudon, G. M. and Koshland, D. E., Jr.,** The chemistry of a reporter group: 2-hydroxy-5-nitrobenzyl bromide, *J. Biol. Chem.,* 245, 2247, 1970.
34. **Riordan, J. R., Sokolovsky, M., and Vallee, B. L.,** Environmentally sensitive tyrosyl residues. Nitration with tetranitromethane, *Biochemistry,* 6, 358, 1967.
35. **Berliner, L. J., ed.,** *Spin Labeling: Theory and Applications,* Academic Press, New York, 1975.
36. **Berliner, L. J. and Wong, S. S.,** Evidence against two "pH locked" conformations of phosphorylated trypsin, *J. Biol. Chem.,* 248, 1118, 1973.
37. **Berliner, L. J. and Wong, S. S.,** Spin-labeled sulfonyl fluorides as active site probes of protease structure. I. Comparison of the active site environments in α-chymotrypsin and trypsin, *J. Biol. Chem.,* 249, 1668, 1974.
38. **Wong, S. S., Quiggle, K., Triplett, C., and Berliner, L. J.,** Spin-labeled sulfonyl fluorides as active site probes of protease structure. II. Spin label synthese and enzyme inhibition, *J. Biol. Chem.,* 249, 1678, 1974.
39. **Berliner, L. J. and Shen, Y. Y.,** Probing active site structure by spin label (ESR) and fluorescence methods, in *Chemistry and Biology of Thrombin,* Lundblad, R. L., Fenton, J. W., II, and Mann, K. G., Eds., Ann Arbor Science, Ann Arbor, Mich., 1977, 197.
40. **Berliner, L. J., Bauer, R. S., Chang, T.-L., Fenton, J. W., II, and Shen, Y. Y. L.,** Active-site topography of human coagulant (α) and noncoagulant (γ) thrombins, *Biochemistry,* 20, 1831, 1981.
41. **Twining, S. S., Sealy, R. C., and Glick, D. M.,** Preparation and activation of spin-labelled pepsinogen, *Biochemistry,* 20, 1267, 1981.
42. **Nesheim, M. E., Kettner, C., Shaw, E., and Mann, K. G.,** Cofactor dependence of factor Xa incorporation into the prothrombinase complex, *J. Biol. Chem.,* 256, 6537, 1981.
43. **Cardin, A. D., Jackson, R. L., and Johnson, J. D.,** 5-Dimethylaminonaphthalene-1-sulfonyl-3-amino-tyrosyl apolipoprotein C-III. Preparation, characterization and interaction with phospholipid vesicles, *J. Biol. Chem.,* 257, 4987, 1982.
44. **Nesheim, M. E., Prendergast, F. G., and Mann, K. G.,** Interactions of a fluorescent active-site-directed inhibitor of thrombin: dansylarginine *N*-(3-ethyl-1,5-pentanediyl) amide, *Biochemistry,* 18, 996, 1979.
45. **Myers, B., II and Glazer, A. N.,** Spectroscopic studies of the exposure of tyrosine residues in proteins with special reference to the subtilisins, *J. Biol. Chem.,* 246, 412, 1971.
46. **Chaiken, I. M. and Smith, E. L.,** Reaction of chloroacetamide with the sulfhydryl groups of papain, *J. Biol. Chem.,* 244, 5087, 1969.

Chapter 2

AMINO ACID ANALYSIS

In the course of exploring some of the early literature on chemical modification, specifically the initial application of photo-oxidation to the study of proteins,[1,2] we had occasion to examine some papers from 1949 to 1952, a little over 30 years ago. We were mildly surprised to see that microbial growth assays and specific enzyme assays were still used to determine changes in the amino acid composition of proteins.[3,4] These studies required substantial quantities of protein and, as such, were limited to proteins such as chymotrypsin, lysozyme, and various albumins which could be readily obtained in such quantities. In addition to the limitations posed by the quantities of material required, the lack of accuracy inherent in these approaches led to the application[5] of chromatography to the problem of determining amino acid composition. This specific era of protein chemistry has been reviewed by Fruton.[6] The application of chromatography to the analysis of amino acid composition occurred in the laboratories of Stanford Moore and William H. Stein at The Rockefeller Institute for Medical Research (now The Rockefeller University). This led to the further development of the ninhydrin reaction for the detection of amino acids[7] and the earlier invention of the automatic fraction collector for use in the chromatographic fractionation. The introduction of ion exchange resin in 1954[8] reduced the time required for the analysis of a single hydrolyzate from 10 days to 5 days. The introduction of the automated amino acid analyzer in 1958[9,10] reduced this time to 24 hr and this in turn had been reduced to 6 hr by 1966 using two-column methodology. The further development of ion-exchange resins made the use of single column analysis possible. This, together with other technical advances, reduced the time required for a single analysis to 2 to 4 hr with sensitivity at the nmol level.[11] Of major importance in this increase in both speed and sensitivity was the development of narrow-bore or microbore columns.[12,13] The critical events in the development of the apparatus used today are also reviewed in several of the preceding references[12,13] and several more recent state-of-the-art reviews.[14-16]

The critical aspects of amino acid analysis at the present time are sample preparation (does the method used to produce amino acids from the peptide or protein under study accurately represent the composition?), preparation of the solvents used for the chromatographic fractionation, and the reproducibility of the analytical system. The third consideration involves the reproducibility of the spectrophotometric analytical system, the maintenance of an extremely reproducible flow rate, and the accuracy of the data analysis system. Implicit in this last area is the accuracy of the amino acid standard used for calibration of the system. With ninhydrin-based detection systems, consideration needs to be given to the possibility of reagent decomposition.

The preparation of the sample will first be discussed. The strategy is somewhat different depending upon the qualitative and quantitative demands on the information desired. If the objective is to rigorously determine the amino acid composition of newly isolated protein, it is essential to know the *chemical* purity of the preparation. Here it would be assumed that a single macromolecular component (protein) is present. In the absence of other analytical information regarding the nature of the protein preparation (i.e., carbohydrate content, nucleic acid content, etc.), it is of critical importance to know precisely how much material (mass) is present at the start so that the investigation can accurately determine what portion (percentage) of the sample is recovered as amino acid. The careful investigator would also know what portion of the sample is moisture or inorganic components detected as % ash. Impurities can also contribute to artifactual results other than inaccurate estimates of the amount of protein present. The presence of nucleic acid in a sample can result in the formation of artifactual amounts of glycine.[17] Inorganic constituents can pose a number of difficulties.

Ionic chromium can irreversibly bind to the ion-exchange columns used for amino acid analysis.[18] This effect can be prevented by addition of one or two extra buffers prior to the increase in pH to an alkaline value (i.e., pH 7.6) for the elution of basic amino acids in the single-column system. One of the authors (RLL) has observed a similar effect of gold. A systematic study has explored the effect of cupric ions and ferric ions[19] on the recovery of amino acids after acid hydrolysis (6N HCl, 20 to 22 hr/110°C). The presence of these metal ions results in the conversion of cysteine/cystine to cysteic acid and methionine to methionine sulfoxide.

In general, adequate results for the amino acid composition of a protein (excluding tryptophan content) can be obtained from careful hydrolysis in 6 N HCl. Effective exclusion of oxygen during the hydrolysis is essential for the accurate determination of composition. This is easily achieved by the careful deaeration of the sample prior to placement in a hydrolysis oven or block.[20] Accurate values for serine, threonine, and tyrosine require samples hydrolyzed for several periods of time (i.e., 22 hr, 48 hr, 72 hr) such that extrapolation back to zero time can be accomplished.

As mentioned above, hydrolysis in 6 N HCl precludes the determination of tryptophan since this amino acid is destroyed under these conditions. Until recently, it was therefore necessary to rely on a spectrophotometric method,[21,22] or somewhat tedious and occasionally unreliable methods[23,24] for alkaline hydrolysis. The introduction of 2-hydroxy-5-nitrobenzyl bromide[25] provided a reliable spectral method but required an independent determination of protein concentration and composition and is only applicable to soluble proteins. The work of Hugli and Moore[26] provided a reliable method for base hydrolysis and subsequent analysis. The conditions developed by these investigators result in the accurate determination of tryptophan, and the presence of carbohydrate does not interfere with the analysis. Duplicate samples of the protein solution (0.100 mℓ in dilute [0.005 M] HCl or NaOH) are placed in a polypropylene centrifuge tube (10.9 × 50 mm) and 25 mg of partially hydrolyzed starch added, followed by 0.50 mℓ 5 N NaOH (freshly prepared). This tube is placed in a glass test tube and deaerated with vacuum in a manner similar to that for acid hydrolysis. Foaming during this procedure can be reduced by the addition of 5 mn 1% octanol (in toluene). This sample is then heated at 110°C or 135°C for the desired period of time. The hydrolyzate is neutralized with HCl and transferred into 0.20 M sodium citrate buffer, pH 4.25. It is noted that the HCl required for neutralization is not added directly to the hydrolyzate but is incorporated in the buffer washes. Analysis is performed immediately or the sample can be frozen at −20°C. Sample buffer with lower pH (i.e., 0.20 M sodium citrate, pH 2.2) should be avoided because of the instability of tryptophan in acidic solution. Although alkaline hydrolysis and spectroscopy are still of value (the use of base hydrolysis, for example, is now not essential for the determination of tryptophan but is essential for the determination of γ-carboxyglutamic acid[27,28]), the work of Liu and co-workers has largely supplanted their use in the quantitative determination of tryptophan in proteins which do not contain substantial carbohydrate. This group reported the development of the use of tosyl acid (p-toluenesulfonic acid) for the hydrolysis of proteins to determine tryptophan.[29] This procedure was effective despite having several disadvantages. The major problem is that tosyl acid is a solid which generally must be recrystallized from HCl prior to use and it is difficult to obtain completely free of HCl. The inclusion of indole or 3-(2-aminoethyl) indole was observed to improve the yield of tryptophan and, therefore, the reproducibility of analyses. The use of 4.0 N methanesulfonic acid as outlined in a subsequent publication from Liu's laboratory[30] has proved to be more effective than tosyl acid. The condition for the hydrolysis of proteins are 4.0 N methanesulfonic acid (containing 0.2% 3-[2-aminoethyl] indole) at 115°C for 22 hr (it may be necessary to include other time periods of hydrolysis such as 48 hr and 72 hr for the reasons given above for 6 N HCl as well as to assure the cleavage of all peptide bonds). Sartin and co-workers[31] have reported that the inclusion of mercaptoacetic acid, phenol, and

3-(2-aminoethyl) indole in 6.0 N HCl (100 $\mu\ell$ sample in H_2O, 100 $\mu\ell$ concentrated HCl, 1 $\mu\ell$ mercaptoacetic acid, 10 $\mu\ell$ of 5% phenol, and 5 $\mu\ell$ of 2% 3-(2-aminoethyl) indole) resulted in the effective recovery of tryptophan for subsequent analysis. The presence of carbohydrate (greater than 5%) results in the excessive loss of tryptophan using any of these procedures such that alkaline hydrolysis or some alternative analytical approach (i.e., 2-hydroxy-5-nitrobenzyl bromide, N-bromosuccinimide) may be preferable.

Most analytical systems are based on the detection of the respective amino acids in the effluent from the analytical columns by reaction with ninhydrin.[7] The ninhydrin systems in use are based either on a methyl cellosolve solvent[7-9] or on dimethyl sulfoxide.[32] Our laboratories use the dimethyl sulfoxide-based system as we find this reagent more stable on storage, the dimethyl sulfoxide does not have the severe toxic properties of methyl cellosolve and the color yields for most amino acids are superior. The primary problem in the use of the ninhydrin system is concerned with reagent preparation. The source of ninhydrin is critical and we have found that only Pierce Chemical Company, Rockford, Ill., has continually provided material of excellence. We have, on occasion, used material from other sources which has had equivalent performance but there was extensive batch-to-batch variation.

The use of fluorogenic reagents to detect amino acids has been increasing during the past decade. Fluorescamine (4-phenylspiro[furan-2-(3H), 1'-phthalan]-3,3'-dione) was introduced by Udenfriend and co-workers[33] for the detection of amino acids. At the time, the reagent provided greater sensitivity for most amino acids than did the ninhydrin reagent but did not detect amino acids such as proline or hydroxyproline. Felix and Turkelsen[34] reported the use of post-column derivatization of proline with N-chlorosuccinimide which permitted effective reaction with fluorescamine. Further studies[35] on the chemistry of the reaction of fluorescamine with amino acids have created a solid base for the use of this reagent for amino acid analysis using colorimetric rather than fluorescence detection. Further developments of note in the use of this reagent include the use of alkaline sodium hypochlorite for the post-column derivatization of amino acids.[36]

Another fluorescence-based system based on the use of o-phthalaldehyde has been developed.[37,38] An excellent review[39] has recently appeared regarding the use of o-phthalaldehyde in amino acid analysis. Our laboratories do not have any practical experience with either fluorescamine or o-phthalaldehyde but it is our impression that o-phthalaldehyde has been more useful. It is water-soluble which makes this compound somewhat easier to use from a technical viewpoint. Both reagents require the post-column derivatization of amino acids which requires a sometime tedious plumbing arrangement. It is also our impression that solvent purity is a more critical factor with either of these reagents than with ninhydrin. The reader is directed to the review[39] cited above for a more critical comparison of the various detection systems.

In conclusion it is important to emphasize that reagent purity and a clean air, clean laboratory environment are both critical for the successful application of current technology in amino acid analysis. A thorough discussion of these factors is beyond the limited scope of this present writing and the reader is directed to several reviews[11,14,15,20,39] which consider these various factors in adequate detail.

REFERENCES

1. **Weil, L. and Buchert, A. R.,** Photooxidation of crystalline β-lactoglobulin in the presence of methylene blue, *Arch. Biochem. Biophys.*, 34, 1, 1951.
2. **Weil, L., James, S., and Buchert, A. R.,** Photooxidation of crystalline chymotrypsin in the presence of methylene blue, *Arch. Biochem. Biophys.*, 46, 266, 1952.

3. **Snell, E. E.,** The microbiological assay of amino acids, *Adv. Protein Chem.,* 2, 85, 1945.
4. **Tristram, G. R.,** The amino acid composition of proteins, in *The Proteins,* Vol. 1, Part A, Neurath, H. and Bailey, K., Eds., Academic Press, New York, 1953, 185.
5. **Moore, S. and Stein, W. H.,** Partition chromatography of amino acids on starch, *Ann. N.Y. Acad. Sci.,* 49, 256, 1948.
6. **Fruton, J. S.,** *Molecules and Life; Historical Essays on the Interplay of Chemistry and Biology,* Wiley-Interscience, New York, 1972, 148.
7. **Moore, S. and Stein, W. H.,** A modified ninhydrin reagent for the photometric determination of amino acids and related compounds, *J. Biol. Chem.,* 211, 907, 1954.
8. **Moore, S. and Stein, W. H.,** Procedures for the chromatographic determination of amino acids on four per cent cross-linked sulfonated polystyrene resins, *J. Biol. Chem.,* 211, 893, 1954.
9. **Moore, S., Spackman, D. H., and Stein, W. H.,** Chromatography of amino acids on sulfonated polystyrene resins. An improved system, *Anal. Chem.,* 30, 1185, 1958.
10. **Spackman, D. H., Stein, W. H., and Moore, S.,** Automatic recording apparatus for use in the chromatography of amino acids, *Anal. Chem.,* 30, 1190, 1958.
11. **Moore, S.,** The precision and sensitivity of amino acid analysis, in *Chemistry and Biology of Peptides,* Meienhofer, J., Ed., Ann Arbor Sci., Ann Arbor, Mich., 1972, 629.
12. **Krejci, V. K. and Machleidt, W.,** Verbessere Methodik der Aminosäuren– analyse im Nanomol-Bereich, *Hoppe Seyler's Z. Physiol. Chem.,* 350, 981, 1969.
13. **Liao, T.-H., Robinson, G. W., and Salnikow, J.,** Use of narrow-bore columns in amino acid analysis, *Anal., Chem.,* 45, 2286, 1973.
14. **Hamilton, P. B.,** Micro and submicro determinations of amino acids by ion-exchange chromatography, *Meth. Enzymol.,* 11, 15, 1967.
15. **Hare, P. E.,** Subnanomole-range amino acid analysis, *Meth. Enzymol.,* 47, 3, 1977.
16. **Benson, J. R.,** Improved ion-exchange resins, *Meth. Enzymol.,* 47, 19, 1977.
17. **Paddock, G. V., Wilson, G. B., and Wang, A.-C.,** Contribution of hydrolyzed nucleic acids and their constituents to the apparent amino acid composition of biological components, *Biochem. Biophys. Res. Commun.,* 87, 946, 1979.
18. **Bech-Anderson, S.,** Single-column analysis of amino acids in hydrolysates of samples containing chromic oxide, *J. Chromatogr.,* 179, 227, 1979.
19. **Brummel, M., Gerbeck, C. M., and Montgomery, R.,** Effects of metals on analytical procedures for amino acids and carbohydrates, *Analyt. Biochem.,* 31, 331, 1969.
20. **Moore, S. and Stein, W. H.,** Chromatographic determination of amino acids by the use of automatic recording equipment, *Meth. Enzymol.,* 6, 819, 1963.
21. **Goodwin, T. W. and Morton, R. A.,** Spectrophotometric determinations of tyrosine and tryptophan in proteins, *Biochem. J.,* 40, 628, 1946.
22. **Edelhoch, H.,** Spectroscopic determination of tryptophan and tyrosine in proteins, *Biochemistry,* 6, 1948, 1967.
23. **Dreze, A.,** Le dosage du tryptophane dans les milieux naturels. II. La stabilité du tryptophane au cours de l'hydrolyse alcaline effectuée en présence d'hydrates de carbone, *Bull. Soc. Chim. Biol.,* 42, 407, 1960.
24. **Noltmann, E. A., Mahowald, T. A., and Kuby, S. A.,** Studies on adenosine triphosphate transphorylases. II. Amino acid composition of adenosine triphosphate-creatine transphosphorylase, *J. Biol. Chem.,* 237, 1146, 1962.
25. **Barman, T. E. and Koshland, D. E., Jr.,** A colorimetric method for the quantitative determination of tryptophan residues in proteins, *J. Biol. Chem.,* 242, 5771, 1967.
26. **Hugli, T. E. and Moore, S.,** Determination of the tryptophan content of proteins by ion-exchange chromatography of alkaline hydrolysates, *J. Biol. Chem.,* 247, 2828, 1972.
27. **Hauschka, P. V.,** Quantitative determination of γ-carboxyglutamic acid in proteins, *Anal. Biochem.,* 80, 212, 1977.
28. **Kuwada, M. and Katayama, K.,** A high-performance liquid chromatographic method for the simultaneous determination of γ-carboxyglutamic acid and glutamic acid in proteins, bone, and urine, *Anal. Biochem.,* 117, 259, 1981.
29. **Liu, T.-Y. and Chang, W. H.,** Hydrolysis of proteins with *p*-toluenesulfonic acid. Determination of tryptophan, *J. Biol. Chem.,* 246, 2842, 1971.
30. **Simpson, R. J., Neuberger, M. R., and Liu, T.-Y.,** Complete amino acid analysis of proteins from a single hydrolysate, *J. Biol. Chem.,* 251, 1936, 1976.
31. **Sartin, J. L., Hugli, T. E., and Liao, T.-H.,** Reactivity of the tryptophan residues in bovine pancreatic deoxyribonuclease with *N*-bromosuccinimide, *J. Biol. Chem.,* 255, 8633, 1980.
32. **Moore, S.,** Amino acid analysis: aqueous dimethyl sulfoxide as solvent for the ninhydrin reaction, *J. Biol. Chem.,* 243, 6281, 1968.

33. **Udenfriend, S., Stein, S., Bohlen, P., and Dairman, W.,** A new fluorometric procedure for assay of amino acids, peptides and proteins in the picomole range, in *Chemistry and Biology of Peptides,* Meienhofer, J., Ed., Ann Arbor Sci., Ann Arbor, Mich., 1972, 655.
34. **Felix, A. M. and Turkelsen, G.,** Total fluorometric amino acid analysis using fluorescamine, *Arch. Biochem. Biophys.,* 157, 177, 1973.
35. **Felix, A. M., Toome, V., DeBernardo, S., and Weigele, M.,** Colorimetric amino acid analysis using fluorescamine, *Arch. Biochem. Biophys.,* 168, 601, 1975.
36. **Böhlen, P., and Mellet, M.,** Automated fluorometric amino acid analysis: the determination of proline and hydroxyproline, *Anal. Biochem.,* 94, 313, 1979.
37. **Cronin, J. R. and Hare, P. E.,** Chromatographic analysis of amino acids and primary amines with *o*-phthalaldehyde detection, *Anal. Biochem.,* 81, 151, 1977.
38. **Lund, E., Thomsen, J., and Brunfeldt, K.,** The use of *o*-phthalaldehyde for fluorescence detection in conventional amino acid analyzers. Sub-nanomole sensitivity in the analysis of phenylthiohydantoin-amino acids, *J. Chromatog.,* 130, 51, 1977.
39. **Lee, K. S. and Drescher, D. G.,** Fluorometric amino-acid analysis with *o*-phthalaldehyde (OPA), *Int. J. Biochem.,* 9, 457, 1978.

Chapter 3

PEPTIDE SEPARATION BY REVERSE-PHASE HIGH PERFORMANCE LIQUID CHROMATOGRAPHY

Many advances in analytical and preparative separation of peptide mixtures arise from development of improved methods and materials for chromatography. The current repertory of methods includes column chromatography in size-exclusion and ion-exchange modes and two-dimensional electrophoresis/chromatography on paper or thin-layer sheets of cellulose or silica. *Methods in Enzymology*, Volumes 11, 25, and 47 review progress in the science and art of these techniques over a 10-year period.[1-11]

Reverse-phase high performance liquid chromatography (HPLC) has developed recently into a useful method for analytical and preparative "mapping" of peptide mixtures, allowing peptides containing sites of chemical modification to be isolated for identification.

Improvement in chromatographic efficiency is achieved by decreasing the particle size of the stationary phase. The resulting decrease in sample bandspreading allows better resolution, faster separation, and higher detection sensitivity. The basis of the technique commonly known as HPLC is a stationary phase made of 10 μm-diameter or smaller particles to yield High Performance. High Pressure is then required to make the liquid mobile phase flow through a column of this packing, and the necessary pump, sample injection valve, detector, and associated equipment lead to a High-Priced system. Snyder and Kirkland[12] provide a thorough discussion of the theory and practice of HPLC.

The original expectation in development of HPLC was that existing thin-layer and column chromatographic separation methods would be transferred to columns packed with small, uniform-sized, rigid particles. This implied a stationary phase of unmodified silica. However, the retention properties of a silica surface are very sensitive to moisture levels. Elution times may be difficult to reproduce, and long column equilibration times may be required following solvent changes. Attempts to improve reproducibility of retention properties by masking the surface silanol groups led to the development of packing materials having covalently bonded surface phases. The most successful of these have been the so-called C-18 or ODS (octa-decylsilyl) packings, which have 18-carbon hydrocarbon chains bonded to the surface silanols. The use of these and other hydrophobic bonded-phase packings is known as "reverse-phase" chromatography because the stationary phase is nonpolar and the mobile phase polar, which happens to be the reverse of most older methods of partition liquid chromatography.

The bonded reverse-phase supports are versatile and provide reproducible retention, rapid equilibration to new solvent conditions, and good column life when used within the pH range of 2 to 7.5. Mobile phases are based on water or aqueous buffers; sample retention is controlled by the proportion of a water-miscible organic solvent, often acetonitrile or an alcohol, in the mobile phase. The higher the volume fraction of organic solvent, the "stronger" the mobile phase, that is, the more hydrophobic and better able it is to elute sample components from the nonpolar stationary phase. Simple mixtures of chemically similar small molecules can be adequately separated in a reasonable time with a single properly selected mobile phase composition ("isocratic" elution). More complex mixtures or those having a wider range of polarities require a solvent compositional gradient for efficient analysis. The retention of peptides and proteins on reverse-phase columns is highly dependent on solvent strength[13] and on their amino acid compositions; gradient elution is necessary for most polypeptide mixtures.

The application of reverse-phase HPLC to proteins and peptides lagged behind its use for smaller molecules. The size and chemical complexity of polypeptides cause their retention characteristics to be affected by a number of factors. These include the pH of the mobile phase, the buffer system used, the organic mobile phase component, the nature of the bonded

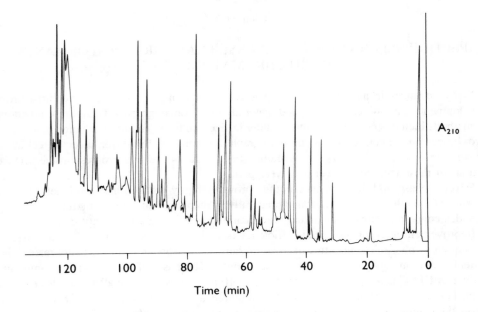

FIGURE 1.   Tryptic digest of reduced and carboxymethylated human blood coagulation Factor IX. Column, Ultrasphere-ODS, 0.46 × 25 cm; flow rate 1 mℓ/min; temperature 25°C. Solvent A = 0.05 *M* sodium phosphate, pH 2.95; solvent B = acetonitrile. Gradient program: 0 to 28% B in 114 min, then to 62% B in 30 min; both segments linear.

stationary phase surface, and the pore size of the stationary phase. Usable elution systems have now been worked out and high-resolution separations can be achieved, particularly of peptides up to about 2 dozen amino acids in length.

For small peptides such as most of those produced by tryptic digestion, ordinary C-18 columns can be used. We have obtained good results[14] with the system recommended by Schroeder and co-workers[15] for the mapping of hemoglobin tryptic peptides.[14,15] A Beckman Ultrasphere-ODS column, 0.46 × 25 cm, is eluted at 1 mℓ/min with a phosphate-acetonitrile gradient at ambient temperature. The aqueous component is approximately 50 m*M* phosphate, pH 2.95. The gradient rate is 0.25% increase in acetonitrile per minute up to 28% acetonitrile, followed by 1.1% increase per minute to a final acetonitrile concentration of 62%. A typical chromatogram of a tryptic digest is shown in Figure 1.

The phosphate-acetonitrile eluent gives good resolution of peptides and allows absorbance monitoring at 210 nm for high-sensitivity detection. Its main defect is a lack of volatility. This level of phosphate salt will interfere in sequence analysis. Schroeder[15] recommends rechromatography of individual peaks using 10 m*M* ammonium acetate adjusted to pH 6.07 with dilute acetic acid as the aqueous component of the mobile phase. We have generally used 5 m*M* phosphate, pH 6.5, for rechromatography, still with an acetonitrile gradient. Up to 1 mℓ of this eluent can be used for sequence analysis without further desalting. In either case the change in pH causes changes in relative peptide mobility so that any peaks which coeluted initially are separated during rechromatography.

Larger peptides give broader peaks[16] and are more sensitive to the nature of both the column packing and the mobile phase. Schroeder[15] notes poor elution of hemoglobin ''core'' peptides in the phosphate-acetonitrile system; they form sharper peaks if oxidized, and also are eluted better from a Zorbax TMS (trimethylsilyl) column than from the Ultrasphere-ODS. Similarly, the broad peak late in the chromatogram shown in Figure 1 is a 35-amino acid glycopeptide (the ''activation peptide'' of human blood coagulation factor IX).[14] To separate it from the phosphate and the coeluting peaks, we chromatographed it on a VydacTP

C-8 column with a gradient from 0.1% trifluoroacetic acid (TFA) to 2-propanol at a flow rate of 1.0 mℓ/min, a gradient rate of 0.5%/min, and a temperature of 45°C.

The Vydac TP packing material has 330 Å-diameter pores. Several laboratories have shown that supports having 300-500 Å pores give better results with large peptides and proteins than the more commonly used supports which have pores of about 100 Å diameter.[17-20]

Either 1-propanol or 2-propanol is a stronger solvent for polypeptides than acetonitrile[21,22] and may be required for elution of some samples not eluted by the latter. However, propanol systems have a higher viscosity than acetonitrile systems. This causes higher column backpressure and lower efficiency (broader peaks).[16]

Phosphate or phosphoric acid has been found to give particularly good results in HPLC of peptides.[14,23-26] It is presumed that ion-pairing between peptide R-NH$_3^+$ groups and the hydrophilic anion H$_2$PO$_4^-$ increase the polarity of the migrating species, thus deceasing their retention. Some workers add perchlorate as a chaotropic agent to phosphate mobile phases.[27-29] Trifluoroacetic acid also works well;[21] larger perfluoridated acids have been used to provide more hydrophobic counterions and thus increase retention of hydrophilic peptides.[18,30,31]

Other mobile phase buffers used in reverse-phase HPLC of polypeptides include triethylamine phosphate[13,32-36] and, as noted previously, ammonium acetate.[14,37-46] All the mobile phase systems mentioned to this point are used with absorbance detection of peptides at 205 to 230 nm (and at 254 or 280 nm or other wavelengths if desired). Fluorescence detection following postcolumn derivatization of peptides[17,28] allows the use of pyridine formate or acetate buffers and is very sensitive but destructive of the portion of the sample diverted to the detector. Wilson and co-workers have compared absorbance and fluorescence detection,[28] finding them both sensitive down to the picomcle level. The fluorescence detector produces some peak broadening from postcolumn mixing of reagents, although this affects only the recorded trace and not the actual resolution of peptides in the portion of the eluent not passed through the detector. They found ultraviolet (UV) absorbance detection more susceptible to impurity and baseline problems.

Chemically modified peptides may be detected by their change in mobility relative to peaks of an unmodified digest, by specific absorbance or fluorescence characteristics of the modifying group and/or by radiolabel incorporated with the modifying group.

Cyanopropyl columns are less retentive than C-18 or C-8 columns and therefore useful for more hydrophobic peptides.[34,41] Alkylphenyl columns have been used for peptide separation.[32,47,48] Newer types of packings include diphenyl and short-chain hydrocarbon bonded phases.

Column performance for polypeptide separations is not necessarily related to performance for separation of small molecules. In the latter situation the sample is considered to undergo a series of equilibration steps between the mobile and stationary phases as it passes through the column.[49] Retention and separation are thus dependent on the column length (number of "theoretical plates" or equilibration steps). With polypeptides this is not the case; 5-cm columns separate as well as 25-cm columns;[19,50] only loading capacity is affected. Pearson et al.[50] attribute this phenomenon to the operation of an adsorption-desorption mechanism instead of the multi-step partitioning which occurs with smaller molecules.

Several groups have determined sets of coefficients for predicting retention of a peptide from its amino acid composition.[16,22,31,51,52] Each set is valid only for a specific chromatographic system (mobile phase, stationary phase, pH, temperature, flow rate, gradient rate, etc.).

REFERENCES

1. **Bennett, J. C.**, Paper chromatography and electrophoresis; special procedures for peptide maps, *Meth. Enzymol.*, 11, 330, 1967.
2. **Hill, R. L. and Delaney, R.**, Peptide mapping with automatic analyzers: use of analyzers and other automatic equipment to monitor peptide separations by column methods, *Meth. Enzymol.*, 11, 339, 1967.
3. **Schroeder, W. A.**, Separation of peptides by chromatography on columns of Dowex 50 with volatile developers, *Meth. Enzymol.*, 11, 351, 1967; 25, 203, 1972.
4. **Schroeder, W. A.**, Separation of peptides by chromatography on columns of Dowex 1 with volatile developers, *Meth. Enzymol.*, 11, 361, 1967; 25, 214, 1972.
5. **Edmundson, A. B.**, Separation of peptides on Aberlite IRC-50, *Meth. Enzymol.*, 11, 369, 1967.
6. **Roy, D. and Konigsberg, W.**, Chromatography of proteins and peptides on diethylaminoethyl cellulose, *Meth. Enzymol.*, 25, 221, 1972.
7. **Gracy, R. W.**, Two-dimensional thin-layer methods, *Meth. Enzymol.*, 47, 195, 1977.
8. **Chin, C. C. Q. and Wold, F.**, Separation of peptides on phosphocellulose and other cellulose ion exchangers, *Meth. Enzymol.*, 47, 204, 1977.
9. **Machleidt, W., Otto, J., and Wachter, E.**, Chromatography on microbore columns, *Meth. Enzymol.*, 47, 210, 1977.
10. **Herman, A. C. and Vanaman, T. C.**, Automated micro procedures for peptide separations, *Meth. Enzymol.*, 47, 220, 1977.
11. **Lai, C. Y.**, Detection of peptides by fluorescence methods, *Meth. Enzymol.*, 47, 236, 1977.
12. **Snyder, L. R. and Kirkland, J. J.**, *Introduction to Modern Liquid Chromatography*, 2nd ed., John Wiley & Sons, New York, 1979.
13. **Wehr, C. T., Correia, L., and Abbott, S. R.**, Evaluation of stationary and mobile phases for reversed-phase high performance liquid chromatography of peptides, *J. Chromatog. Sci.*, 20, 114, 1982.
14. **Noyes, C. M. and Lundblad, R. L.**, unpublished data, 1981.
15. **Schroeder, W. A., Shelton, J. B., and Shelton, J. R.**, High performance liquid chromatography in the identification of human hemoglobin variants, in *Advances in Hemoglobin Analysis*, Hanash, S. M. and Brewer, G. J., Eds., Alan R. Liss, New York, 1981, 1.
16. **Meek, J. L. and Rossetti, Z. L.**, Factors affecting retention and resolution of peptides in high-performance liquid chromatography, *J. Chromatog.*, 211, 15, 1981.
17. **Lewis, R. V., Fallon, A., Stein, S., Gibson, K. D., and Udenfriend, S.**, Supports for reverse-phase high-performance liquid chromatography of large proteins, *Anal. Biochem.*, 104, 153, 1980.
18. **van der Rest, M., Bennett, H. P. J., Solomon, S., and Glorieux, F. H.**, Separation of collagen cyanogen bromide-derived peptides by reversed-phase high-performance liquid chromatography, *Biochem. J.*, 191, 253, 1980.
19. **Pearson, J. D., Mahoney, W. C., Hermodson, M. A., and Regnier, F. E.**, Reversed-phase supports for the resolution of large denatured protein fragments, *J. Chromatog.*, 207, 325, 1981.
20. **Wilson, K. J., Van Wieringen, E., Klauser, S., Berchtold, M. W., and Hughes, G. J.**, Comparison of the high-performance liquid chromatography of peptides and proteins on 100- and 300-Å reversed-phase supports, *J. Chromatog.*, 237, 407, 1982.
21. **Mahoney, W. C. and Hermodson, M. A.**, Separation of large denatured peptides by reverse phase high performance liquid chromatography. Trifluoroacetic acid as a peptide solvent, *J. Biol. Chem.*, 255, 11199, 1980.
22. **Wilson, K. J., Honegger, A., Stötzel, R. R., and Hughes, G. J.**, The behaviour of peptides on reverse-phase supports during high-pressure liquid chromatography, *Biochem. J.*, 199, 31, 1981.
23. **Hancock, W. S., Bishop, C. A., Prestidge, R. L., Harding, D. R. K., and Hearn, M. T. W.**, High-pressure liquid chromatography of peptides and proteins. II. The use of phosphoric acid in the analysis of underivatised peptides by reversed-phase high-pressure liquid chromatography, *J. Chromatog.*, 153, 391, 1978.
24. **Fullmer, C. S. and Wasserman, R. H.**, Analytical peptide mapping by high performance liquid chromatography. Application to intestinal calcium-binding proteins, *J. Biol. Chem.*, 254, 7208, 1979.
25. **O'Hare, M. J. and Nice, E. C.**, Hydrophobic high-performance liquid chromatography of hormonal polypeptides and proteins on alkylsilane-bonded silica, *J. Chromatog.*, 171, 209, 1979.
26. **Nice, E. C. and O'Hare, M. J.**, Simultaneous separation of β-lipotrophin, adrenocorticotropic hormone, endorphins and enkephalins by high-performance liquid chromatography, *J. Chromatog.*, 162, 401, 1979.
27. **Meek, J. L.**, Prediction of peptide retention times in high-pressure liquid chromatography on the basis of amino acid composition, *Proc. Natl. Acad. Sci. U.S.A.*, 77, 1632, 1980.
28. **Wilson, K. J., Honegger, A., and Hughes, G. J.**, Comparison of buffers and detection systems for high-pressure liquid chromatography of peptide mixtures, *Biochem. J.*, 199, 43, 1981.

29. **Strydom, D. J. and Vallee, B. L.,** Characterization of human alcohol dehydrogenase isoenzymes by high-performance liquid chromatographic peptide mapping, *Anal. Biochem.,* 123, 422, 1982.

30. **Bennett, H. J. P., Browne, C. A., and Solomon, S.,** The use of perfluorinated carboxylic acids in the reversed-phase HPLC of peptides, *J. Liquid Chromatog.,* 3, 1353, 1980.

31. **Browne, C. A., Bennett, H. J. P., and Solomon, S.,** The isolation of peptides by high-performance liquid chromatography using predicted elution positions, *Anal. Biochem.,* 124, 201, 1982.

32. **Rivier, J. E.,** Use of trialkyl ammonium phosphate (TAAP) buffers in reverse phase HPLC for high resolution and high recovery of peptides and proteins, *J. Liquid Chromatog.,* 1, 343, 1978.

33. **Biedermann, K., Montali, U., Martin, B., Svendsen, I., and Ottesen, M.,** The amino acid sequence of proteinase A inhibitor 3 from baker's yeast, *Carlsberg Res. Comm.,* 45, 225, 1980.

34. **Chaiken, I. M. and Hough, C. J.,** Mapping and isolation of large peptide fragments from bovine neurophysins and biosynthetic neurophysin-containing species by high-performance liquid chromatography, *Anal. Biochem.,* 107, 11, 1980.

35. **Hancock, W. S., Capra, J. D., Bradley, W. A., and Sparrow, J. T.,** The use of reversed-phase high-performance liquid chromatography with radial compression for the analysis of peptide and protein mixtures, *J. Chromatog.,* 206, 59, 1981.

36. **Hearn, M. T. W. and Grego, B.,** High-performance liquid chromatography of amino acids, peptides and proteins. XXXVI. Organic solvent modifier effects in the separation of unprotected peptides by reversed-phase liquid chromatography, *J. Chromatog.,* 218, 497, 1981.

37. **Wilson, J. B., Lam, H., Pravatmuang, P., and Huisman, T. H. J..** Separation of tryptic peptides of normal and abnormal α, β, γ, and δ hemoglobin chains by high-performance liquid chromatography, *J. Chromatog.,* 179, 271, 1979.

38. **Hancock, W. S. and Sparrow, J. T.,** Use of mixed-mode, high-performance liquid chromatography for the separation of peptide and protein mixtures, *J. Chromatog.,* 206, 71, 1981.

39. **Kehl, M. and Henschen, A.,** Characterization of the peptides released at the fibrinogen-fibrin conversion using high performance liquid chromatography, in *High Performance Chromatography in Protein and Peptide Chemistry,* Lottspeich, F., Henschen, A., and Hupe, K.-P., Eds., Walter de Gruyter, Berlin, 1981, 339.

40. **Kratzin, H. and Yang, C.-Y.,** Separation of enzymatic hydrolysates by reverse-phase HPLC, in *High Performance Chromatography in Protein and Peptide Chemistry,* Lottspeich, F., Henschen, A., and Hupe, K.-P., Eds., Walter de Gruyter, Berlin, 1981, 269.

41. **Ponstingl, H., Krauhs, E., Little, M., Kempf, T., Hofer-Warbinek, R., and Ade, W.,** Elucidation of tubulin amino acid sequence: preparative separation of peptides by reversed phase HPLC, in *High Performance Chromatography in Protein and Peptide Chemistry,* Lottspeich, F., Henschen, A., and Hupe, K.-P., Eds., Walter de Gruyter, Berlin, 1981, 325.

42. **Stüber, K., and Beyreuther, K.,** Preparative separation of peptides by high performance liquid chromatography. Use of volatile buffers transparent in the ultraviolet, in *High Performance Chromatography in Protein and Peptide Chemistry,* Lottspeich, F., Henschen, A., and Hupe, K.-P., Eds., Walter de Gruyter, Berlin, 1981, 205.

43. **Yang, C.-Y. and Kratzin, H.,** Chromatography and rechromatography of peptide mixtures by reverse-phase HPLC, in *High Performance Chromatography in Protein and Peptide Chemistry,* Lottspeich, F., Henschen, A., and Hupe, K.-P., Eds., Walter de Gruyter, Berlin, 1981, 283.

44. **Anderson, J. K. and Mole, J. E.,** Adaptation of reverse-phase high-performance liquid chromatography for the isolation and sequence analysis of peptides from plasma amyloid P-component, *Anal. Biochem.,* 123, 413, 1982.

45. **Lambert, D. T., Stachelek, C., Varga, J. M., and Lerner, A. B.,** Iodination of β-melanotropin, *J. Biol. Chem.,* 257, 8211, 1982.

46. **Oray, B., Jahani, M., and Gracy, R. W.,** High-sensitivity peptide mapping of triosephosphate isomerase: a comparison of high-performance liquid chromatography with two-dimensional thin-layer methods, *Anal. Biochem.,* 125, 131, 1982.

47. **Hancock, W. S., Bishop, C. A., Meyer, L. J., Harding, D. R. K., and Hearn, M. T. W.,** High-pressure liquid chromatography of peptides and proteins. VI. Rapid analysis of peptides by high-pressure liquid chromatography with hydrophobic ion-pairing of amino groups, *J. Chromatog.,* 161, 291, 1978.

48. **Hancock, W. S., Bishop, C. A., Battersby, J. E., Harding, D. R. K., and Hearn, M. T. W.,** High-pressure liquid chromatography of peptides and proteins. XI. The use of cationic reagents for the analysis of peptides by high-pressure liquid chromatography, *J. Chromatog.,* 168, 377, 1979.

49. **Martin, A. J. P. and Synge, R. L. M.,** A new form of chromatogram employing two liquid phases. I. A theory of chromatography. II. Application to the micro-determination of the higher monoamino-acids in proteins, *Biochem. J.,* 35, 1358, 1941.

50. **Pearson, J. D., Lin, N. T., and Regnier, F. E.,** The importance of silica type for reverse-phase protein separations, *Anal. Biochem.,* 124, 217, 1982.

51. **Su, S.-J., Grego, B., Niven, B., and Hearn, M. T. W.,** Analysis of group retention contributions for peptides separated by reversed phase high performance liquid chromatography, *J. Liquid Chromatog.*, 4, 1745, 1981.
52. **Sasagawa, T., Okuyama, T., and Teller, D. C.,** Prediction of peptide retention times in reversed-phase high-performance liquid chromatography during linear gradient elution, *J. Chromatog.*, 240, 329, 1982.

Chapter 4

METHODS FOR SEQUENCE DETERMINATION

TABLE OF CONTENTS

I. INTRODUCTION

Determination of the amino acid sequence around a site of chemical modification provides unequivocal identification of its location in the proteins, providing the complete sequence of the protein is known. Although analysis of amino acid composition alone may allow identification of the peptide containing the modification, even a few cycles of sequence analysis serve to verify its purity and identity.

Most methods for the determination of amino acid sequence use the degradation scheme developed by Edman.[1,2] (The procedure was originally described as a micro-method requiring "only" 10 mg of amino acid per cycle.) Phenylisothiocyanate (PITC) reacts with the NH_2-terminal amino acid of a polypeptide at basic pH to form a phenylthiocarbamyl derivative (reaction 1). Acidification, generally with anhydrous acid, then cleaves off the first amino acid as its 2-anilino-5-thiazolinone derivative and exposes the amino group of the second amino acid (reaction 2). The derivatized amino acid is removed by extraction, the remaining polypeptide is dried, an the cycle is repeated. The large difference in pH between coupling and cleavage allows the degradation to proceed by discrete steps, unlike exopeptidase digestions, which begin to release the second amino acid before all of the first has been removed.

Procedures for sequence analysis by the Edman degradation fall into two major categories differing in the type of method used for determination of the amino acid at each step. In so-called indirect Edman procedures, a portion of the sample is removed at the beginning and after each cycle and analyzed either quantitatively for total amino acid composition or qualitatively for the identity of the NH_2-terminal. In direct methods, the thiazolinone produced at each step is extracted, converted to the more stable thiohydantoin (reaction 3), and identified. Manual procedures of both types are used; automated sequenators have generally used direct identification.

The following points apply to all Edman degradation techniques:

1. All procedures and identification methods should be practiced first on known peptides.
2. Ideally, known examples of derivatives from all amino acids from actual degradations should be observed in the identification system to be used.
3. Purity of reagents and solvents, particularly freedom from aldehydes and oxidants, is important.
4. Degradations are performed in an atmosphere of nitrogen or argon to exclude oxygen and air pollutants that can cause side reactions and block further degradation.[3,4] Manual degradation may be carried out in a glove box[5] or closed reaction vessel[6] to facilitate maintenance of an inert atmosphere.

II. INDIRECT EDMAN DEGRADATION

A. Subtractive

An aliquot of the peptide is removed after each cycle and subjected to hydrolysis and quantitative amino acid analysis. A typical procedure is outlined by Konigsberg:[7]

The medium for coupling can be (1) 50% aqueous pyridine containing 2% triethylamine, (2) *N*-ethylmorpholine per acetic acid per 95% ethanol per water (60:1.5:500:438), or (3) 50% pyridine containing 5% dimethylallylamine. The peptide is dissolved in coupling buffer and treated with a 50-fold excess of PITC at 37°C for 2 hr. (Exclusion of oxygen and contaminants during coupling is particularly important in the subtractive method, since partial blockage of peptide will cause nonintegral loss of amino acid.)

After coupling, the sample is evaporated nearly to dryness and extracted three times with 1 to 2 mℓ benzene, then dried. Cleavage is carried out in anhydrous trifluoroacetic acid (TFA) at 25°C for 1 hr or 40°C for 15 min. The TFA is evaporated; the residue is dissolved in 0.2 *M* acetic acid and heated at 40°C for 10 min. It is then extracted three times with benzene. An appropriate aliquot of the acetic acid solution is removed for hydrolysis and amino acid analysis (see Chapter 2). The remaining sample is dried and the cycle is repeated.

Subtractive degradation is applicable only to peptides small enough that loss of a single amino acid is clearly distinguishable in the amino acid analysis. The amount of peptide

PITC

+ H₂NCHR₁CONHCHR₂CO ~~~~~

peptide

(1)

OH⁻ | COUPLING

S
‖
N – C – NHCHR₁CONHCHR₂CO ~~~~~

phenylthiocarbamyl peptide

(2)

H+ | CLEAVAGE

+ H₂NCHR₂CO ~~~~~

shortened
peptide

2 – anilino – s – thiazolinone

aq. H+ | CONVERSION

(3)

3 – phenyl – 2 – thiohydantoin

needed depends on the sensitivity of the analyzer and the number of Edman cycles to be carried out. Glu/Gln and Asp/Asn are not differentiated by the usual acid hydrolysis procedure, since Gln and Asn are deamidated.

B. Dansyl-Edman

Intensely fluorescent derivatives result from the reaction of 1-dimethylaminonaphthalene-5-sulfonyl (dansyl) chloride with free amino groups. Dansyl amino acid derivatives are stable to acid hydrolysis and exhibit yellow fluorescence.[8] The NH$_2$-terminal of a peptide can therefore be identified by reacting it with dansyl chloride, hydrolyzing the peptide, and identifying the fluorescent derivative by electrophoresis or chromatography.[8-11] For sequence determination, an aliquot is removed at each cycle of degradation and analyzed in this way. Since 10 pmol of dansyl amino acid is readily detected,[11] the amount removed can be less than in subtractive Edman degradation. Also, the sample need only be washed after cleavage, so extractive losses of peptide may be less than with other procedures. The method requires no major equipment and is considered simple and easily learned.[12]

Instructions for dansyl-Edman degradation of about a nanomole of peptide are given by Bruton and Hartley.[11] The sample is dissolved in 20 μℓ of water, a 10 pmol aliquot is removed for dansylation, and 20 μℓ of 5% (v/v) PITC in pyridine is added to the remainder. The tube is flushed with nitrogen and heated at 45°C for 1 hr, then dried in a vacuum desiccator at 60°C for 30 min. TFA (20 μℓ) is added and the sample is incubated at 45°C for 30 min and dried *in vacuo* over NaOH. The residue is dissolved in 25 μℓ of water and extracted with four 150-μℓ portions of *n*-butyl acetate. The aqueous solution is dried *in vacuo* and redissolved in 20 μℓ of water. Another 10-pmol aliquot is removed for dansylation and the cycle repeated on the remainder.

The aliquots removed at each cycle are dried in a vacuum desiccator over P$_2$O$_5$. Two successive 2-μℓ portions of 0.1 *M* NaHCO$_3$ are added and dried to remove ammonia. Dansyl chloride solution (1 μℓ of a 1:1 mixture of water and a solution of 2.5 mg dansyl chloride/mℓ acetone) is added and the sample incubated at 37°C for 1 hr. After drying, 5 μℓ of 6 *N* HCl is added. The tube is sealed and the peptide is hydrolyzed at 105°C for 16 hr. The dansyl amino acids are identified by 2-dimensional thin layer chromatography on 5 × 5 cm polyamide sheets.[10] A fine capillary is used for sample application to keep the spot as small as possible. A mixture of standards is run on the reverse side of the sheet. The solvent for the first dimension is 1.5% (v/v) formic acid. In the second dimension benzene/acetic acid 9:1 (or less toxic toluene/acetic acid 10:1) is run, followed in the same direction by ethyl acetate/methanol/acetic acid 20:1:1. The plate is dried and examined under UV light after each step. Resolution of certain spots may require a fourth solvent, 0.05 *M* trisodium phosphate/ethanol 3:1, also run in the second direction. As in the subtractive method, glutamine and asparagine are deamidated during hydrolysis. The thiazolinones may be recovered from the organic solvent, converted to the phenylthiohydantoins, and identified by one of the methods described later.

Gray and Smith developed a rapid dansyl-Edman procedure for five or six cycles on small peptides.[13] The sample is divided at the start into portions corresponding to the number of cycles to be carried out, and one tube is set aside before each cycle. The desired number of cycles is performed, with drying at 70°C *in vacuo* after each coupling and each cleavage, but with no solvent wash until the end. At that time all tubes are washed with water-saturated ethyl acetate to remove phenylthiourea and diphenylthiourea. Dansylation, hydrolysis, and identification are performed as described above.

Detailed instructions for a modified dansyl-Edman procedure suitable for analysis of proteins (which tend to become insoluble in the solvents used for peptides) are given by Weiner et al.[14]

III. DIRECT EDMAN DEGRADATION

A. Manual PITC

An extensive discussion of the factors affecting the Edman chemistry and of possible reagent, solvent, and temperature options for manual degradation is given by Tarr.[6] The manual Edman procedure described by Peterson et al.[15] has been widely used.

Levy recommends the following technique for 1 to 10 nmol of peptide.[16] Acid-washed polypropylene microcentrifuge tubes are used, and all manipulations of liquids are done with polypropylene micropipet tips. Argon is preferred for flushing since it is heavier than nitrogen and thus stays in the tubes better. The buffer for coupling consists of 15 mℓ pyridine, 1.18 mℓ dimethylallylamine, and 10 mℓ water; the pH is adjusted to 9.5 with TFA. Norleucine, 25 to 50 nmol, is added at the beginning of each cycle as a carrier and internal standard.

Coupling takes place with 40 $\mu\ell$ buffer plus 3 $\mu\ell$ PITC for 30 min at 50°C under argon. The sample is washed twice with heptane/ethyl acetate 10:1 and once with heptane/ethyl acetate 2:1, with vigorous mixing followed by centrifugation to separate the phases. The sample (aqueous phase) is dried under vacuum. Cleavage follows in 20 $\mu\ell$ TFA for 20 min at 50°C. After vacuum drying the sample is dissolved in 40 $\mu\ell$ of 30% pyridine, extracted with three 150-$\mu\ell$ portions of benzene:ethyl acetate (1/2) and the cycle is repeated.

The benzene/ethyl acetate extract containing the thiazolinone is dried. Conversion to the phenylthiohydantoin is done in 40 $\mu\ell$ 1 *N* HCl for 10 min at 80°C, followed by extraction with three 50-$\mu\ell$ portions of ethyl acetate. All PTHs except PTH-Arg, -His, and -CysSO$_3$H are extracted by the ethyl acetate. The PTHs are identified by HPLC or one of the other methods described in Section IV. of this chapter.

B. Manual DABITC/PITC

The Edman degradation can be carried out with the compound dimethylaminoazobenzene-4'-isothiocyanate (DABITC) in place of PITC.[17,18] The thiohydantoin amino acid derivatives (DABTHs) of this compound are red (ϵ_M, 436 nm $\cong$ 34,000), so they can be detected visually on thin layer sheets, with the further advantage that derivatives of non-amino acid impurities appear blue or are invisible. However, quantitative coupling with DABITC requires a temperature of 75°C. Consequently a double coupling is performed instead at 52°C; the second coupling is with PITC to drive the reaction to completion.

In the method developed by Chang et al.,[18] 2 to 8 nmol of peptide or protein are dried in an acid-washed tube and dissolved in 80 μℓ of 50% pyridine. Freshly prepared DABITC solution (2.82 mg/mℓ pyridine), 40 μℓ is added. The tube is flushed with nitrogen and heated at 52°C for 50 min. PITC, 10 μℓ, is added, followed by another 30 min incubation at 52°C. The mixture is extracted two or three times by 0.5 mℓ portions of heptane/ethyl acetate 2:1, with vortexing followed by centrifugation. The aqueous phase is dried under high vacuum. The sample is dissolved in 50 μℓ anhydrous TFA, flushed with nitrogen, sealed with a glass stopper, and heated at 52°C for 15 min. After drying in a vacuum desiccator, it is dissolved in 50 μℓ of water and extracted with 200 μℓ of butyl acetate. The sample is dried and the cycle repeated.

For conversion, the butyl acetate extract is evaporated; 20 μℓ of water and 40 μℓ of acetic acid saturated with HCl are added. Conversion is at 52°C for 50 min. The sample is dried, redissolved in ethanol, and applied to polyamide sheets for thin layer chromatography. Acetic acid/water, 1:2, is run in the first dimension and toluene/n-hexane/acetic acid, 2:1:1, in the second. DABTH-Leu and -Ile are resolved on silica gel plates in chloroform/ethanol, 100:3. The plates are dried and exposed to HCl vapor; DABTH spots are red. Sensitivity of detection is 5 to 25 pmol on polyamide, 25 to 50 pmol on silica.

Wilson, Hughes, and co-workers[19,20] have adapted the DABITC/PITC degradation scheme for use in automated sequenators. They prefer 1N HCl or 20% TFA for the conversion step. DABITC is light-sensitive and not very stable in solution, so appropriate precautions must be taken.

C. Spinning-Cup Sequenator

In 1967 Edman and Begg[21] published a description of a machine for automatically performing the Edman degradation of proteins. Essentially a "robot chemist", the sequenator or sequencer carries out the coupling, wash, cleavage, and extraction steps of the cycle. Conversion of thiazolinones to thiohydantoins is still often performed manually, although automated converters have been devised.[22,23]

The heart of the sequenator is a spinning glass cup in a temperature-controlled reaction chamber. The protein or peptide sample is deposited in a thin film in the lower half of the cup. Reagents for coupling or cleavage are added to cover the film. Solvents for extraction are run into the cup for several minutes, so that the liquid flows up over the precipitated sample, out through a tube at the top of the cup, and is directed by a valve to waste or a fraction collector. The sample can be dried by a stream of nitrogen followed by progressive degrees of vacuum. An inert atmosphere is maintained in the reaction chamber.

In addition to the convenience of unattended operation, the sequenator gives better results than manual degradation in extended sequence analyses of intact proteins. These tend to become insoluble precipitates after a few cycles of manual degradation, with concomitant poor yields and out-of-phase results. The spinning cup of the sequenator keeps the sample spread out in a thin film, thus maintaining efficiency of reaction and extraction.

For small peptides, on the other hand, the sequenator does not have as clear a performance advantage over manual techniques. With these samples it is more difficult to achieve adequate extraction of reagents, by-products, and thiazolinones without also washing out the sample. Manual procedures offer more flexibility and better control. A great deal of effort has gone into improving sequenator retention of small peptides, including modifying them to decrease their solubility in organic solvents[26,27] and use of nondegradable or artificial carriers.[28,29] The most successful carrier has been the polymer 1,5-dimethyl-1,5-diazaundecamethylene polymethobromide, commonly known as Polybrene.[30,31] Addition of 1 to 3 mg of Polybrene to the cup with the sample allows complete sequence analysis of even subnanomolar amounts of many peptides. Extremely hydrophobic peptides may still require manual degradation, but in general, automated sequence analysis is now as feasible for peptides as for proteins.

As with all sequence analysis techniques, the amount of sample required for automated Edman degradation is determined by the sensitivity of the identification method used, the repetitive yield on successive cycles of degradation, and the number of cycles to be run. Identification depends on both absolute sensitivity and levels of background contaminants. Besides whatever background is obtained from reagent impurities and side reactions, an increasing level of amino acid background arises during a sequence analysis because of small amounts of nonspecific cleavage of peptide bonds.[21] This background is proportional to the size of the polypeptide, and runs on intact proteins, commonly limited in length by the decreasing signal-to-noise ratio arising from decreasing yield and increasing background from nonspecific cleavage and cycle-to-cycle overlap. Peptides up to a few dozen residues in length, on the other hand, produce very little cumulative background and can be followed to the limits imposed by the sensitivity of the detection method used and/or the background from the degradation procedure.

We are able to achieve 93% repetitive yield on 250 pmol starting yield of a peptide using a unmodified Beckman 890C sequenator, Polybrene carrier, and Beckman chemicals. Identification of PTHs by HPLC with this system is possible down to 10 to 50 pmol, so complete sequence analysis of small peptides is possible and at least 20 cycles on larger ones. The first amino acids to be unidentifiable tend to be serine and proline.

We use the 0.1 *M* Quadrol sequenator program of Brauer et al.[30] with the following modifications. After coupling, the buffer is flushed from the delivery line with a short delivery of ethyl acetate. Delivery of heptafluorobutyric acid (HFBA) for cleavage is modified to use a minimum amount of acid, to avoid blowing acid vapors through the effluent valve, and to provide thorough flushing of acid from the delivery line. HFBA is delivered for 3 seconds, followed by 20 seconds of restricted vacuum to draw it into the cup. After cleavage and a preliminary drying step, *n*-chlorobutane is delivered briefly through the same line and dried. Extraction with *n*-chlorobutane follows. Three mg of Polybrene is added to the cup and put through one complete cycle before adding the sample. (Some authors find more extensive precycling of Polybrene necessary,[31,32] so there may be lot-to-lot variations in its purity).

Some laboratories have devised extensive modifications of the spinning-cup sequenator.[31,33,34] These changes in conjunction with meticulous purification of reagents provide decreased background levels and improved repetitive yields, thus allowing extended degradation on subnanomolar quantities of sample. Applications related to the topic of this book should not generally require such measures.

Automated devices for the conversion of thiazolinones to PTHs have been described.[22,23] Although they increase convenience and possibly recovery levels of unstable derivatives, their use places more exacting demands on system cleanliness, reagent purity, and the resolution of the identification method, and may in turn necessitate the further sequenator and reagent improvements described below. With an automatic converter the conversion mixture is dried down in a single phase. Thus, all the PTHs must be distinguished from each other and from all by-products during identification, whereas manual conversion divides the products between an organic and an aqueous phase, which are analyzed separately.

D. Solid-Phase Sequenator

One approach to the problem of sample loss during washing is covalent attachment of the peptide to a solid support.[35] The support beads are then placed in a temperature-controlled column or reaction chamber, and reagents and solvents are passed through.

Methods of covalent attachment have been summarized by Laursen.[36] Peptides can be attached to appropriate supports through the COOH-terminal homoserine produced by cyanogen bromide digestion, through lysine ϵ-NH$_2$ groups, or through carboxyl groups. In the latter method, aspartic and glutamic acid side chains tend to become bound in addition to the COOH-terminal carboxyl, causing gaps at those points in the sequence. Support materials are derivatized glass or polystyrene beads.[35,37]

The necessary repertoire of materials and attachment methods is cumbersome but may be used to achieve separation of peptide mixtures by selective coupling and sequences of COOH-terminal portions of larger peptides by trypsin digestion of bound material.[36]

Commercially available equipment for automated solid-phase Edman degradation is less expensive than liquid-phase (spinning-cup) machines. It may be possible to use less expensive chemicals, since more extensive washing is possible. Disadvantages are the extra time and work required for sample attachment to the solid support and sample losses due to incomplete coupling.

Powers[38] describes solid-phase methods for spinning-cup sequenators. These include both placing beads with peptides attached in the sequenator cup (and hoping they won't wash out and plug the effluent valve) and attaching peptides directly to the glass wall of the cup. These techniques have been mostly superseded by the simpler use of Polybrene carrier, but may be useful for particular problems.

Chang[39] has performed manual sequence analysis by the DABITC/PITC procedure on peptides attached to glass beads. No advantage was found over the corresponding liquid-phase manual method except for small peptides where extractive loss would be a problem without attachment.

E. Gas-Liquid Solid Phase Sequenator

A new type of automatic sequenator has been devised by Hewick et al.[40] and is now commercially available. The instrument is designed specifically for sequence analysis of small amounts (<10 nmol) of proteins and peptides; information has been obtained on as little as 5 pmol. The sample is deposited on a Polybrene-impregnated glass fiber filter and placed in a 50-$\mu\ell$ reaction chamber. Buffer for coupling and acid for cleavage are delivered as vapors in a stream of argon; other reagents and solvents are delivered as liquids.

The miniaturized design reduces reagent consumption; however, extremely pure (and therefore expensive in purchase cost or in time) chemicals are necessary for analysis at the low picomole level.

IV. IDENTIFICATION METHODS FOR DIRECT EDMAN DEGRADATION

A. High Performance Liquid Chromatography

HPLC offers high resolution, quantitative analysis, and high sensitivity; it is currently the method of choice for identification of PTH amino acids in sequence analysis.

Several dozen publications have appeared since 1973 on separation of PTHs by HPLC.[41] The methods most commonly used today are similar in general principle to that of Zimmerman et al.;[42] i.e., a 25-cm C-18 reverse-phase column is used, with isocratic or gradient elution by a mobile phase of buffered aqueous sodium acetate and acetonitrile at elevated temperature. Other bonded-phase column packings such as cyanopropyl[43] or phenylalkyl[44] types are also used, as are other solvents and buffer salts. Even C-18 columns from different manufacturers differ in retention characteristics, so experimental conditions optimized for one column may have to be adjusted if a column from another source is used.

The desired result is adequate separation of all PTH amino acids from each other and from any by-products of the Edman degradation. If the conversion procedure used includes extraction from an acidic aqueous phase by ethyl acetate, then PTH-Arg, -His and hydrophilic by-products remain in the aqueous phase and need only be separated from each other, which is not difficult. Retention of the PTHs extracted by ethyl acetate is adjusted by changing the temperature and the proportion of organic solvent in the mobile phase.

We find it helpful in establishing optimal conditions to measure the retention of the individual PTHs over a range of solvent strengths and temperatures. If retention times are plotted on a logarithmic scale as a function of solvent strength or temperature, separation between lines for the individual PTHs is essentially proportional to resolution.[45]

All the PTHs except PTH-dehydrothreonine (derived from PTH-Thr) are easily detected at 254 nm with the common mercury lamp absorbance monitor. PTH-dehydrothreonine is detectable at 313 or 323 nm. In the case of serine, little or no PTH-Ser is seen in samples from our sequencing system (manual conversion in 1 N HCl) nor is there any peak visible at 323 nm; the only peak seen elutes after all the other PTHs and absorbs at 254 nm. It may be a dehydrated and polymerized product, as described by Chang for the DABITC degradation.[46] Other laboratories do report detection of serine as PTH-dehydroserine absorbing at 313 to 323 nm and eluting at an intermediate position among the PTHs[34,47] or (in the presence of dithiothreitol) as a derivative eluting near PTH-Ala and absorbing at 254 nm.[31,48] It is, therefore, essential that elution and absorbance characteristics be determined for the product(s) of Edman degradation of known serine residues in the particular sequence determination procedure to be used.

Gradient elution is more rapid than isocratic but may not save time in repetitive analyses because of the time required for column re-equilibration. It also causes baseline fluctuations at high detector sensitivity. The principal advantage of gradient elution in PTH analysis may be increased sensitivity of detection for the late-eluting serine derivative mentioned above. Reproducibility and flexibility are important in a solvent gradient system for PTH analysis. We use a complex gradient profile which includes a nominal temporary drop in solvent strength at one point (although the actual change is damped by the mixing chamber and is more likely only a plateau, see Figure 1). Total analysis time including re-equilibration is under 35 min. A faster analysis is possible in which PTH-Lys elutes between PTH-Trp and -Phe, but background peaks (e.g., diphenylthiourea) from the sequenator then co-elute with PTH-Trp.

Detection of PTHs is possible at the picomole level with the UV monitor. As discussed earlier (Section III.C.), sensitivity in actual sequence analysis depends partly on levels of background from the sample and the degradation system.

B. Thin Layer Chromatography

Thin layer methods are inexpensive and have the particular advantage of speed, since derivatives from a number of cycles of degradation can be processed simultaneously. Lack of quantitation is less of a problem in sequence analysis of pure samples of peptides, since they do not develop the degree of overlap and background from nonspecific cleavage that long runs on proteins produce. PTHs are detected under 254 nm UV light as dark spots on a fluorescent background.

PTH identification procedures using silica gel plates include those of Jeppsson and Sjöquist,[49] Solal and Bernard,[50] and Inagami and Murakami.[51] We have found that a change in the binder used in Eastman plates in the late 1970s made them incompatible with some of the organic solvents recommended in these papers. However, the solvent xylene/95% ethanol/acetic acid, 50:50:0.5, recommended by Inagami[52] for PTH-His and -Arg is compatible with the new binder, and the same mixture with twice the proportion of xylene works well for most of the other PTHs. Inagami and Murakami[51] recommend marking the positions of spots under UV light, then spraying the plate with 0.5% ninhydrin in *n*-butanol, drying, and heating at 95 to 110°C for 10 to 15 min. Many of the PTHs develop characteristic colors useful for their identification. Sensitivity is about 1 to 5 nmol.

Kulbe[53] recommends solvent systems for 1- and 2-dimensional TLC on polyamide sheets.

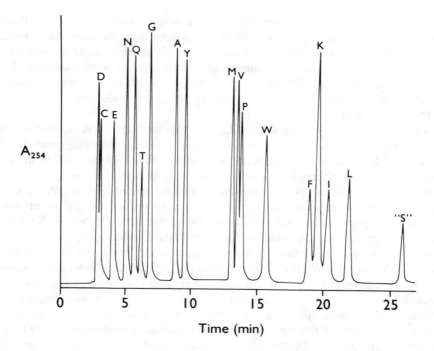

FIGURE 1. Separation of PTH amino acids on Ultrasphere-ODS column, 0.46 × 25 cm. Flow rate 1 mℓ/min, temperature 55°C. Solvent A = 0.01 *M* sodium acetate, pH 4.9; solvent B = acetonitrile. Gradient program: percent B increased linearly from 27 to 46.9% in 11 min, then dropped immediately to 30% and held there 1 min, and finally increased to 49% in 10 min. PTHs: D, aspartic acid; C, carboxymethylcysteine; E, glutamic acid; N, asparagine; Q, glutamine; T, threonine; G, glycine; A, alanine; Y, tyrosine; M, methionine; V, valine; P, proline; W, tryptophan; F, phenylalanine; K, lysine; I, isoleucine; L, leucine; "S", serine product.

Solvent I is toluene/*n*-pentane/acetic acid, 60:30:16, and Solvent II is 25% aqueous acetic acid. Solvent II is run at right angles to Solvent I for maximum resolution, or in the same direction if several sequence cycles are to be analyzed on one sheet. Sensitivity is 0.05 to 0.2 nmol.

C. Gas-Liquid Chromatography

Although purchase of a gas chromatograph for use in sequence analysis can no longer be recommended since HPLC gives better results for a similar investment, some users may have access to an instrument. It was the primary means of PTH identification in many sequence laboratories until the rise of HPLC. The equipment is reliable and the results reproducible. Instrumental requirements are a gas chromatograph with flame ionization detector, heated glass-lined injection port, and temperature programming capability to at least 300°C. Deficiencies of gas chromatography for PTH analysis include difficulties with the more polar PTHs and lack of detector selectivity.

The PTHs of the polar amino acids are seen poorly or not at all because of their lack of volatility and/or interactions with the coated silica stationary phase. Modification by trimethylsilylation improves their detection, but PTH-His and -Arg are still not generally detectable, and recoveries of acids and amides may be low. PTH-Ser is not readily distinguished from PTH-carboxymethyl cysteine, and PTH-Thr may be confused with PTH-Pro. The result of these problems is a substantial use of secondary identification methods such as TLC or hydrolysis plus amino acid analysis, particularly in the determination of unknown sequence.

The flame ionization detector generally used is very sensitive but detects all organic compounds. Thus, all volatile by-products and inpurities from the degradation procedure will be detected. This background plus baseline rise during the temperature program limit the practical use of the instrument to samples of about 5 nmol or more. Detectors of higher specificity might have alleviated this latter problem, but the other problems would have remained. HPLC has replaced gas chromatography because it gives good results with more of the PTH amino acids.

The small number of publications on GLC methods for PTHs is in striking contrast to the number on HPLC. The experimental variables are fewer, consisting primarily of the composition of the liquid stationary phase used to coat the silica particles of the column packing. Of the commonly available silicone coatings, only those containing a chlorophenyl group have adequate separating ability for some of the closely eluting PTHs. Pisano et al.[54,55] favored stationary mixtures including the chlorophenyl silicones DC-560 or SP-400 (Supelco), but most protein sequence laboratories found that 10% DC-560 or SP-400 alone gave adequate separation, faster analysis times, and fewer problems.

SP-400 is available only on precoated support and should be ordered as 10% SP-400 on Chromosorb W (HP or AW-DMCS) or Supelcoport, 100 to 120 mesh. Those wishing to coat their own support can obtain DC-560 or GE Versilube F-50 (Analabs). Sildanized glass columns, 4 ft $\times$ 2 mm I.D., are used and can easily be packed by applying mild suction to one end of the column (plugged with silanized glass wool) and tapping gently as the packing material is poured in through a funnel. The column is conditioned at 300°C or more with helium carrier gas flowing through it before connecting the outlet end to the detector.

Temperature program and carrier gas flow are adjusted to give adequate separation of PTHs in a reasonable analysis time, generally about 15 min plus time for temperature re-equilibration between samples. Silylation of PTHs can be performed by simultaneous injection of equal volumes of sample solution and *N,O-bis*(trimethylsilyl)acetamide.

D. Hydrolysis and Amino Acid Analysis

This procedure is slower than others, a disadvantage for its use as the primary means of identification in extended sequence analysis. However, it does provide quantitative results and is valuable as a supplementary or confirmatory means of identification.

Smithies et al.[56] recommend hydrolysis of the thiazolinones or PTHs in 57% hydriodic acid (HI) at 127°C for 20 hr. PTH-alanine, -serine, -carboxymethylcysteine or -cysteine all hydrolyze to alanine. Threonine is identified as α-aminobutyric acid. PTH-tryptophan gives glycine plus alanine, and methionine is destroyed. Alkaline hydrolysis in 0.2 *M* NaOH plus 0.1 *M* sodium dithionite allows recovery of methionine and tryptophan and differentiation of alanine from serine or cysteine.

Mendez and Lai[57] prefer hydrolysis in 5.7 *N* HCl containing 0.1% $SnCl_2$ for 4 hr at 150°C.

V. COOH-TERMINAL DEGRADATION

Methods for sequential degradation from the carboxyl terminals of peptides have not yet achieved the efficiency of the NH_2-terminal methods just described. Most such procedures involve reaction with thiocyanate or thiocyanic acid to form the 2-thiohydantoin derivative of the COOH-terminal amino acid.[58] A recent paper by Meuth et al.[59] describing recent advances in technique states that repetitive yield is about 90%. Attachment of the sample to a derivatized glass support aids in separating peptide from reagents but seems to lower recoveries of some amino acid thiohydantoins.

REFERENCES

1. **Edman, P.,** A method for the determination of the amino acid sequence in peptides, *Arch. Biochem.,* 22, 475, 1949.
2. **Edman, P.,** Method for determination of the amino acid sequence in peptides, *Acta Chem. Scand.,* 4, 283, 1950.
3. **Ilse, D. and Edman, P.,** The formation of 3-phenyl-2-thiohydantoins from phenylthiohydantoin amino acids, *Australian J. Chem.,* 16, 411, 1963.
4. **Schroeder, W. A.,** Degradation of peptides by the Edman method with direct identification of the PTH-amino acid, *Meth. Enzymol.,* 11, 445, 1967.
5. **Meagher, R. B.,** Rapid manual sequencing of multiple peptide samples in a nitrogen chamber, *Anal. Biochem.,* 67, 404, 1975.
6. **Tarr, G.,** Improved manual sequencing methods, *Meth. Enzymol.,* 47, 335, 1977.
7. **Konigsberg, W.,** Subtractive Edman degradation, *Meth. Enzymol.,* 11, 461, 1967.
8. **Gray, W. R. and Hartley, B. S.,** A fluorescent end-group reagent for proteins and peptides, *Biochem. J.,* 89, 59P, 1963.
9. **Gray, W. R.,** Sequential degradation plus dansylation, *Meth. Enzymol.,* 11, 469, 1967.
10. **Hartley, B. S.,** Strategy and tactics in protein chemistry, *Biochem. J.,* 119, 805, 1970.
11. **Bruton, C. J. and Hartley, B. S.,** Chemical studies on methionyl-tRNA synthetase from *Escherichia coli, J. Mol. Biol.,* 52, 165, 1970.
12. **Croft, L. R.,** *Introduction to Protein Sequence Analysis,* John Wiley & Sons, New York, 1980, 110.
13. **Gray, W. R. and Smith, J. F.,** Rapid sequence analysis of small peptides, *Anal. Biochem.,* 33, 36, 1970.
14. **Weiner, A. M., Platt, T., and Weber, K.,** Amino-terminal sequence analysis of proteins purified on a nanomole scale by gel electrophoresis, *J. Biol. Chem.,* 247, 3242, 1972.
15. **Peterson, J. D., Nehrlich, S., Oyer, P. E., and Steiner, D. F.,** Determination of the amino acid sequence of the monkey, sheep and dog pro-insulin C-peptides by a semi-micro Edman degradation procedure, *J. Biol. Chem.,* 247, 4866, 1972.
16. **Levy, W.,** Manual Edman sequencing techniques for proteins and peptides at the nanomole level, *Meth. Enzymol.,* 79, 27, 1981.
17. **Chang, J. Y., Creaser, E. H., and Bentley, K. W.,** 4-*N*,*N*-Dimethylamino-azobenzene 4′-isothiocyanate, a new chromophoric reagent for protein sequence analysis, *Biochem. J.,* 153, 607, 1976.
18. **Chang, J. Y., Brauer, D., and Wittmann-Liebold, B.,** Micro-sequence analysis of peptides and proteins using 4-*N*,*N*-dimethylaminoazobenzene 4′-isothiocyanate/phenylisothiocyanate double coupling method, *FEBS Lett.,* 93, 205, 1978.
19. **Hughes, G. J., Winterhalter, K. H., Lutz, H., and Wilson, K. J.,** Microsequence analysis. III. Automatic solid-phase sequencing using DABITC, *FEBS Lett.,* 108, 92, 1979.
20. **Wilson, K. J., Hunziker, P., and Hughes, G. J.,** Microsequence analysis. IV. Automatic liquid-phase sequencing using DABITC, *FEBS Lett.,* 108, 98, 1979.
21. **Edman, P. and Begg, G.,** A protein sequenator, *Eur. J. Biochem.,* 1, 80, 1967.
22. **Wittmann-Liebold, B., Graffunder, H., and Kohls, H.,** A device coupled to a modified sequenator for the automated conversion of anilinothiazolinones into PTH amino acids, *Anal. Biochem.,* 75, 621, 1976.
23. **Horn, M. J. and Bonner, A. G.,** Automatic conversion, microsequencing and other advances in solid-phase sequence analysis, in *Solid Phase Methods in Protein Sequence Analysis,* Previero, A. and Coletti-Previero, M.-A., Eds., Elsevier/North Holland, Amsterdam, 1977, 163.
24. **Foster, J. A., Bruenger, E., Hu, C. L., Albertson, K., and Franzblau, C.,** A new improved technique for automated sequencing of non-polar peptides, *Biochem. Biophys. Res. Commun.,* 53, 70, 1973.
25. **Braunitzer, G., Schrank, B., and Ruhfus, A.,** Versüche zum vollständigen und automatischen Abbau von Peptiden nach der Quadrolmethod, *Hoppe-Seyler's Z. Physiol. Chem.,* 351, 1589, 1970.
26. **Rochat, H., Bechis, G., Kopeyan, C., Gregoire, J., and Van Rietschoten, J.,** Use of parvalbumin as a protecting protein in the sequenator: an easy and efficient way for sequencing small amounts of peptides, *FEBS Lett.,* 64, 404, 1976.
27. **Silver, J. and Hood, L.,** Automated microsequence analysis in the presence of a synthetic carrier, *Anal. Biochem.,* 60, 285, 1974.
28. **Tarr, G. E., Beecher, J. F., Bell, M., and McKean, D. J.,** Polyquaternary amines prevent peptide loss from sequenators, *Anal. Biochem.,* 84, 622, 1978.
29. **Klapper, D. G., Wilde, C. E., III, and Capra, J. D.,** Automated amino sequence analysis of small peptides utilizing Polybrene, *Anal. Biochem.,* 85, 126, 1978.
30. **Brauer, A. W., Margolies, M. N., and Haber, E.,** The application of 0.1 *M* Quadrol to the microsequence of proteins and the sequence of tryptic peptides, *Biochemistry,* 14, 3029, 1975.
31. **Hunkapiller, M. W. and Hood, L. E.,** Direct microsequence analysis of polypeptides using an improved sequenator, a nonprotein carrier (Polybrene), and high pressure liquid chromatography, *Biochemistry,* 17, 2124, 1978.

32. **Henschen-Edman, A. and Lottspeich, F.**, Aspects on automated and microscale sequencing, in *Methods in Peptide and Protein Sequence Analysis,* Birr, C., Ed., Elsevier/North-Holland, Amsterdam, 1980, 105.

33. **Wittmann-Liebold, B. and Lehmann, A.**, New approaches to sequencing by micro- and automatic solid phase technique, in *Methods in Peptide and Protein Sequence Analysis,* Birr, C., Ed., Elsevier/North Holland, Amsterdam, 1980, 49.

34. **Shively, J. E.**, Sequence determinations of proteins and peptides at the nanomole and subnanomole level with a modified spinning cup sequenator, *Meth. Enzymol.,* 79, 31, 1981.

35. **Laursen, R. A.**, Solid-phase Edman degradation. An automatic peptide sequencer, *Eur. J. Biochem.,* 20, 89, 1971.

36. **Laursen, R. A.**, Coupling techniques in solid-phase sequencing, *Meth. Enzymol.,* 47, 277, 1977.

37. **Machleidt, W. and Wachter, E.**, New supports in solid-phase sequencing, *Meth. Enzymol.,* 47, 263, 1977.

38. **Powers, D. A.**, Solid-phase sequencing in spinning-cup sequenators, *Meth. Enzymol.,* 47, 299, 1977.

39. **Chang, J. Y.**, Manual solid phase sequence analysis of polypeptides using 4-N,N-dimethylaminoazobenzene 4'-isothiocyanate, *Biochim. Biophys. Acta,* 578, 188, 1979.

40. **Hewick, R. M., Hunkapiller, M. W., Hood, L. E., and Dreyer, W. J.**, A gas-liquid solid phase peptide and protein sequenator, *J. Biol. Chem.,* 256, 7990, 1981.

41. **Anon.**, *Applications of the Beckman Protein Peptide Sequencer, a Bibliography,* Beckman Instruments, Palo Alto, 1980, 7.

42. **Zimmerman, C. L., Appella, E., and Pisano, J. J.**, Rapid analysis of amino acid phenylthiohydantoins by high-performance liquid chromatography, *Anal. Biochem.,* 77, 569, 1977.

43. **Johnson, N. D., Hunkapiller, M. W., and Hood, L. E.**, Analysis of phenylthiohydantoin amino acids by high-performance liquid chromatography on DuPont Zorbax cyanopropylsilane columns, *Anal. Biochem.,* 100, 335, 1979.

44. **Henderson, L. E., Copeland, T. D., and Oroszlan, S.**, Separation of amino acid phenylthiohydantoins by high-performance liquid chromatography on phenylalkyl support, *Anal. Biochem.,* 102, 1, 1980.

45. **Noyes, C. M.**, Optimization of complex separations in HPLC: application to phenylthiohydantoin amino acids, *J. Chromatogr.,* 226, 451, 1983.

46. **Chang, J. Y.**, The destruction of serine and threonine thiohydantoins during the sequence determination of peptides by 4-N,N-dimethylaminoazobenzene 4'-isothiocyanate, *Biochim. Biophys. Acta,* 578, 175, 1979.

47. **Wittmann-Liebold, B.**, Application of HPLC-techniques to the separation of PTH-amino acid derivatives and peptides, in *High Performance Chromatography in Protein and Peptide Chemistry,* Lottspeich, F., Henschen, A., and Hupe, K.-P., Eds., Walter de Gruyter, Berlin, 1981, 223.

48. **Shively, J. E., Hawke, D., and Jones, B. N.**, Microsequence analysis of peptides and proteins. III. Artifacts and the effects of impurities on analysis, *Anal. Biochem.,* 120, 312, 1982.

49. **Jeppsson, J.-O. and Sjöquist, J.**, Thin-layer chromatography of PTH amino acids, *Anal. Biochem.,* 18, 264, 1967.

50. **Solal, M. C. and Bernard, J. L.**, Miniature thin-layer chromatography of phenylthiohydantoin amino acids. Application of automatic Edman degradation, *J. Chromatogr.,* 80, 140, 1973.

51. **Inagami, T. and Murakami, K.**, Identification of phenylthiohydantoin amino acids by thin-layer chromatography on a plastic-backed silica-gel plate, *Anal. Biochem.,* 47, 501, 1972.

52. **Inagami, T.**, Simultaneous identification of PTH derivatives of histidine and arginine by thin-layer chromatography, *Anal. Biochem.,* 52, 318, 1973.

53. **Kulbe, K. D.**, Micropolyamide thin-layer chromatography of phenylthiohydantoin amino acids (PTH) at the subnanomolar level. A rapid microtechnique for simultaneous multisample identification after automated Edman degradation, *Anal. Biochem.,* 59, 564, 1974.

54. **Pisano, J. J. and Bronzert, T. J.**, Analysis of amino acid phenylthiohydantoins by gas chromatography, *J. Biol. Chem.,* 244, 5597, 1969.

55. **Pisano, J. J., Bronzert, T. J., and Brewer, H. G., Jr.**, Advances in the gas chromatographic analysis of amino acid phenyl- and methylthiohydantoins, *Anal. Biochem.,* 45, 43, 1972.

56. **Smithies, O., Gibson, D., Fanning, E. M., Goodfliesh, R. M., Gilman, J. G., and Ballantyne, D. L.**, Quantitative procedures for use with the Edman-Begg sequenator. Partial sequences of two unusual immunoglobulin light chains, Rzf and Sac, *Biochemistry,* 10, 4912, 1971.

57. **Mendez, E. and Lai, C. Y.**, Regeneration of amino acids from thiazolinones formed in the Edman degradation, *Anal. Biochem.,* 68, 47, 1975.

58. **Stark, G. S.**, Sequential degradation of peptides from their carboxyl termini with ammonium thiocyanate and acetic anhydride, *Biochemistry,* 7, 1796, 1968.

59. **Meuth, J. L., Harris, D. E., Dwulet, F. E., Crowl-Powers, M. L., and Gurd, F. R. N.**, Stepwise sequence determination from the carboxyl terminus of peptides, *Biochemistry,* 21, 3750, 1982.

Chapter 5

CHEMICAL CLEAVAGE OF PEPTIDE BONDS

The elucidation of the covalent structure of any biopolymer consisting of nonidentical monomer units requires the development of specific, reproducible methods for the cleavage of the biopolymer into fragments. Proteolytic enzymes such as trypsin, chymotrypsin and pepsin have proved quite useful in the cleavage of specific peptide bonds in proteins. However, the nature of certain amino acid residues has permitted the development of nonenzymatic chemical methods for the cleavage of certain peptide bonds.

We would guess that partial acid hydrolysis is the most ancient of the various chemical approaches to the cleavage of specific peptide bonds. This was one of the principal approaches used by Sanger in his primary structure work on insulin.[1-3] Although there are problems with this approach reflecting the variability in the yields of specific fragments as well as danger of deamidation of certain asparagine and glutamine residues, it has continued to prove useful in the structural analysis of proteins. Schultz[4] has reviewed the recent work in this area. The general principle of partial acid hydrolysis is based on the use of dilute acid at a pH just adequate to maintain the β-carboxyl group of aspartic acid in the protonated form. Under these conditions, peptide bonds in which the carboxyl moeity is contributed by aspartic acid are cleaved 100-fold more rapidly than other peptide bonds. Specifically, the use of 0.03 N HCl *in vacuo* at 105°C for 20 hr has been found to be satisfactory. This process has also been recently studied by Tsung and Fraenkel-Conrat.[5]

Cleavage of methionine-containing peptide bonds by cyanogen bromide[6] is certainly the most widely used method for specific chemical cleavage of peptide bonds. The reaction cleaves peptide bonds in which methionine contributes the carboxyl moeity. Methionine is converted into homoserine lactone during this process. There are several reasons for this high degree of popularity. First, the reaction is reasonably quantitative although, as indicated below, variable amounts of cyanogen bromide (CNBr) might be required. Second, the methionine content of most proteins is low[7] enough that a reasonably small number of fragments are obtained, providing a distinct advantage in primary structure analysis. Finally, the knowledge of the primary structure around methionine residues is of particular value in the design of primary DNA probes for the isolation and characterization of cDNA fragments in recombinant DNA research. The chemistry of this reaction is straightforward, involving the nucleophilic attack of the thioether sulfur on the carbon in cyanogen bromide followed by cyclization to form the iminolactone, which is hydrolyzed by water resulting in cleavage of the peptide bond. At acid pH this reaction does not generally, in and by itself, affect any other amino acid with the exception of cysteine, which is converted to cysteic acid. In this regard it is noted that one would rarely be working with a protein or peptide containing free sulfhydryl groups. The yield of cleavage is the sum of homoserine and homoserine lactone after acid hydrolysis. This value is probably best determined by allowing complete conversion to homoserine with base at room temperature. The CNBr reaction proceeds best in weak acid. Early work used 0.1 M HCl while more recent studies have used 70% formic acid. The use of formic acid, however, has been recently shown to result on occasion in the blockage of amino terminal residues which frustrates attempts at subsequent primary structure analysis.[8] The majority of work in our laboratory has used 70% formic acid at room temperature for 24 to 48 hr with an estimated 20- to 100-fold molar excess of cyanogen bromide (added as a solid to the protein or peptide dissolved in formic acid). However, an examination of the effect of solutions of formic acid on amino terminal availability coupled with an increasing experience on our part of apparently blocked amino-terminal amino acids has prompted us to perform cyanogen bromide cleavages in 0.1 M HCl. Solutions of acetic acid could also be used as solvent for the cyanogen bromide reaction. The molar ratio of cyanogen bromide to methionyl residues needs to be established for each peptide and protein under study. In general we use a tenfold molar excess of cyanogen bromide. (This is added as a solid in most instances although we have used an aqueous solution on occasion). It should be recognized that a suitable molar excess will have to be established for each protein. In the work on the structure of the pancreatic deoxyribonuclease it was necessary to use a 3000-fold molar excess to cleave a particular methionine-serine peptide bond.[9] In this regard it is of interest that methionine-123 in human serum albumin is converted to homoserine lactone by treatment with cyanogen bromide without concomitant peptide bond cleavage.[10] The conversion of methionine to methionine sulfoxide under conditions used for the cyanogen bromide cleavage has been reported.[11] With a tenfold molar excess of cyanogen bromide for 22 hr at ambient temperature, 1% conversion to methionine sulfoxide was observed in 70% formic acid, 8% conversion in 0.1 M HCl, 64% conversion in 0.1 M citrate, pH 3.5 and 97% conversion in 0.1 M phosphate, pH 6.5.

The other methods for chemical cleavage of specific peptide bonds have been used somewhat infrequently. This is, in part, a reflection of the considerable success experienced with the cyanogen bromide reaction as well as the increased ability to determine more primary structure during a single run with the improved automated Edman degradation.

A number of methods have been proposed for chemical cleavage at cysteine residues. One approach is based on the conversion of cysteine to dehydroalanine[12-14] and subsequent hydrolysis with either acid or base to release pyruvic acid and involves the conversion of cysteine to the dialkyl sulfonic salt with methyl bromide or methyl iodide at pH 6.0 and subsequent β-elimination in dilute bicarbonate with mild heating.[14] The use of 2,4-dinitrofluorobenzene for the modification of cysteine to form the *S*-dinitrophenyl derivatives at pH 5.6 has been reported.[14] The β-elimination of these derivatives was accomplished with sodium methoxide in methanol. Cleavage of the dehydroalanine-containing peptide bond was accomplished by heating (100°C) in dilute acid (0.01 M HCl) for 1 hr. This reaction mixture was then lyophilized, treated with a volume of 0.1 M NaOH equivalent to the original volume of acid and one fifth volume of 30% hydrogen peroxide and then heated at 37°C for 30 min. The reaction mixture was then neutralized with acetic acid and excess peroxide removed with catalase. Alternatively, cleavage can be accomplished with bromide or performic acid.

The cleavage of peptide bonds containing cystine (disulfide groups) has been examined in some detail[15] although, as with cleavage at dehydroalanine derived from cysteine or serine as described above, wide application of this approach has not been achieved. In this reaction, cyanide reacts with cystine to yield a sulfhydryl and a thiocyano group. The thiocyano-containing derivative at pH less than 8 will cyclize to form an acyliminothiazolidine ring which will then undergo hydrolysis to cleave the cystine peptide bond. The formation of the iminothiazolidine can be followed by absorbance at 235 nm. An application of this approach has been advanced by Jacobson et al.[16] *S*-cyanocysteine is obtained by reaction of cysteine or cystine with 2-nitro-5-thiocyano-benzoic acid. Cleavage of the *S*-cyanocysteine is achieved by incubation in 0.1 M sodium borate, 6 M guanidine, pH 9.0 at 37°C with the formation of 2-iminothiazolidine-4-carboxyl peptides. Virtually 100% cleavage was achieved for several proteins. This results in the formation of a free carboxyl group and a ''blocked'' amino terminal peptide (2-iminothiazolidine-4-carboxyl) derivative. Schaffer and Stark[17] have proposed a catalyst prepared from nickel chloride and sodium borohydride for the conversion of the 2-iminothiazolidine-4-carboxylate to alanine. These investigators also noted that cleavage could occur at phenylalanyl-seryl and phenylalanyl-threonyl peptide bonds. Lu and Gracy[18] have employed 2-nitro-5-thiocyanobenzoic acid to convert the cysteinyl residues in human placental glucosephosphate isomerase to *S*-cyanocysteine, followed by cleavage at the modified cysteine residues. Conversion to the *S*-cyanocysteinyl derivative was accomplished with a five to tenfold molar excess of 2-nitro-5-thiocyanobenzoic acid in

0.2 *M* Tris-acetate, *6M* guanidinium chloride, pH 9.0 (protein previously incubated with adequate dithiothreitol — fourfold molar excess over sulfhydryls in the protein) for 5 hr at 37°C. The modified protein was dialyzed extensively against 10% acetic acid and lyophilized. Cleavage of *S*-cyanylated protein was achieved by incubation in 0.2 *M* Tris-acetate, 6 *M* guanidium chloride, pH 9.0 at 37°C for 2 hr. The average extent of cleavage obtained was approximately 80%.

Specific cleavage at tryptophanyl residues in peptides and proteins has been frequently used to obtain specific fragments. Cleavage of tryptophanyl peptide bonds with *N*-bromo-succinimide can occur as a side reaction of the *N*-bromosuccinimide oxidation of tryptophanyl residues (see Volume II, Chapter 2) but generally requires a substantial molar excess of reagent with mild acid.[19] This reaction is generally accomplished in 70% acetic acid. Although cleavage is generally restricted to tryptophanyl residues, cleavage can also occur at tyrosyl and histidinyl residues. Cleavage at tryptophanyl residues under the above conditions generally occurs with an efficiency of 50 to 80% with peptides but is substantially less with proteins (10 to 50%).

BNPS-Skatole (2-(2-nitrophenylsulfenyl)-3-methyl-3-bromoindolenine) has been used for the cleavage of peptide bonds involving tryptophan.[20] The reaction conditions are similar to those utilized for *N*-bromosuccinimide, and the reaction mechanism is similar in terms of the production of an "active bromide". It is reported to be somewhat more selective than *N*-bromosuccinimide, but nonspecific cleavages do occur as does the conversion of methionine to methionine sulfoxide. The yield of peptide bond cleavage is similar to that reported with *N*-bromosuccinimide.

The use of 2,4,6-tribromo-4-methyl-cyclohexadione (TBC) for the cleavage of trypto-phanyl peptide bonds in proteins has been advanced by Burstein and Patchornik.[21] The cleavage is fairly specific for tryptophanyl residues but modification of other amino acid residues was noted (tyrosine, methionine, cysteine, etc). Optimal conditions for the cleavage reaction were a threefold excess of reagent at pH 3.0 at ambient conditions for 15 min. Generally, 60 to 80% acetic acid is used as the solvent and the reaction is allowed to proceed in the dark. Approximately 50% cleavage of tryptophanyl-containing peptide bonds is obtained with synthetic peptides such as *N*-benzyloxy-carbonyl-tryptophanyl-glycine while 5 to 60% yields are reported with proteins such as lysozyme.

More recently, the specific cleavage of tryptophanyl peptide bonds with *N*-chlorosuccinimide[22] fragment has been reported. This is far more specific than the reagents described above. The cleavage of tryptophanyl peptide bonds requires a twofold excess of *N*-chlorosuccinimide in 50% acetic acid under ambient conditions. Cleavage of other peptide bonds was not detected under these conditions but methionine is converted to methionine sulfoxide, and cysteine to cystine. Model peptides were cleaved in approximately 40% yield while with several proteins yields from 19 to 50% were reported. Mechanistically, the reaction proceeds as described above for *N*-bromosuccinimide. The *N*-chlorosuccinimide should be recrystallized from ethyl acetate prior to use. Similar cleavage of tryptophanyl peptide bonds occurs during chemical and peroxidase-catalyzed iodination.[23] This includes the use of chloramine T and lactoperoxidase-catalyzed iodination. This reaction occurs optimally at pH 5.0.

The cleavage of protein at asparaginyl-glycyl peptide bonds with hydroxylamine[24] has proved useful in selected circumstances. The reaction is generally performed in the presence of 6 *M* guanidium chloride at pH 9.0 with 2 *M* NH_2OH. The pH of the solution is maintained either with a pH-stat or with 0.2 *M* potassium carbonate. Generally, as with other means of peptide bond cleavage, optimal results are obtained with the reduced and alkylated protein. The reaction will yield a new amino-terminal amino acid and aspartyl hydroxyamate.

Cleavage at peptide bonds where the carboxyl group is contributed by tryptophan occurs upon reaction with *o*-iodosobenzoic acid. The reaction has been studied in detail by Hermodson and co-workers.[25,26] The reaction can be reasonably specific for tryptophan although some modification of methionine to form methionine sulfoxide is observed. The reaction is performed in 60 to 80% acetic acid in the presence of a denaturing agent such as guanidine. The occasional modification of tyrosyl residues seen with some preparations of *o*-iodoso-benzoic acid has been shown to be a property of *o*-iodoxybenzoic acid contamination of certain *o*-iodosobenzoic acid preparations.[26] Pretreatment of the *o*-iodosobenzoic acid preparations with *p*-cresol obviates cleavage at tyrosyl peptide bonds.

REFERENCES

1. **Sanger, F.,** The terminal peptides of insulin, *Biochem. J.,* 45, 563, 1949.
2. **Sanger, F. and Tuppy, H.,** The amino-acid sequence in the phenylalanyl chain of insulin. I. The identification of lower peptides from partial hydrolysates, *Biochem. J.,* 49, 463, 1951.
3. **Sanger, F. and Thompson, E. O. P.,** The amino acid sequence in the glycyl chain of insulin. I. The identification of lower peptides from partial hydrolysates, *Biochem. J.,* 53, 353, 1953.
4. **Schultz, J.,** Cleavage at aspartic acid, *Meth. Enzymol.,* 11, 255, 1967.
5. **Tsung, C. M. and Fraenkel-Conrat, H.,** Preferential release of aspartic acid by dilute acid treatment of tryptic peptides, *Biochemistry,* 4, 793, 1965.
6. **Gross, E.,** The cyanogen bromide reaction, *Meth. Enzymol.,* 11, 238, 1967.
7. **Tristram, G. R. and Smith, R. H.,** Amino acid composition of certain proteins, in *The Proteins,* 2nd ed., Neurath, H., Ed., Academic Press, New York, 1963, 45.
8. **Shively, J. E., Hawke, D., and Jones, B. N.,** Microsequence analysis of peptides and proteins. III. Artifacts and the effects of impurities on analysis, *Analyt. Biochem.,* 120, 312, 1982.
9. **Liao, T.-H., Salnikow, J., Moore, S., and Stein, W. H.,** Bovine pancreatic deoxyribonuclease A. Isolation of cyanogen bromide peptides; complete covalent structure of the polypeptide chain, *J. Biol. Chem.,* 248, 1489, 1973.
10. **Doyen, N. and LaPresle, C.,** Partial non-cleavage by cyanogen bromide of a methionine-cystine bond from human serum albumin and bovine α-lactalbumin, *Biochem. J.,* 177, 251, 1979.
11. **Joppich-Kuhn, R., Corkill, J. A., and Giese, R. W.,** Oxidation of methionine to methionine sulfoxide as a side reaction of cyanogen bromide cleavage, *Analyt. Biochem.,* 119, 73, 1982.
12. **Witkop, B. and Ramachandran, L. K.,** Progress in non-enzymatic selective modification and cleavage of proteins, *Metabolism,* 13, 1016, 1964.
13. **Patchornik, A. and Sokolovsky, M.,** Nonenzymatic cleavages of peptide chains at the cysteine and serine residues through their conversion into dehydroalanine. I. Hydrolytic and oxidative cleavage of dehydroalanine residues, *J. Am. Chem. Soc.,* 86, 1206, 1964.
14. **Sokolovsky, M., Sadeh, T., and Patchornik, A.,** Nonenzymatic cleavages of peptide chains at the cysteine and serine residues though their conversion to dehydroalanine (DHAL). II. The specific chemical cleavage of cysteinyl peptides, *J. Am. Chem. Soc.,* 86, 1212, 1964.
15. **Catsimpoolas, N. and Wood, J. L.,** Specific cleavage of cystine peptides by cyanide, *J. Biol. Chem.,* 241, 1790, 1966.
16. **Jacobson, G. R., Schaffer, M. H., Stark, G. R., and Vanaman, T. C.,** Specific chemical cleavage in high yield at the amino peptide bonds of cysteine and cystine residues, *J. Biol. Chem.,* 248, 6583, 1973.
17. **Schaffer, M. H. and Stark, G. R.,** Ring cleavage of 2-iminothiazolidine-4-carboxylates by catalytic reduction. A potential method for unblocking peptides formed by specific chemical cleavage at half-cystine residues, *Biochem. Biophys. Res. Commun.,* 71, 1040, 1976.
18. **Lu, H. S. and Gracy, R. W.,** Specific cleavage of glucosephosphate isomerase at cysteinyl residues using 2-nitro-5-thiocyanobenzoic acid: analyses of peptides eluted from polyacrylamide gels and localization of active site histidyl and lysyl residues, *Arch. Biochem. Biophys.,* 212, 347, 1981.
19. **Ramachandran, L. K. and Witkop, B.,** N-Bromosuccinimide cleavage of peptides, *Meth. Enzymol.,* 11, 283, 1967.
20. **Fontana, A.,** Modification of tryptophan with BNPS-skatole (2-(2-nitrophenylsulfenyl)-3-methyl-3-bromoindolenine), *Meth. Enzymol.,* 25, 419, 1972.
21. **Burstein, Y. and Patchornik, A.,** Selective chemical cleavage of tryptophanyl peptide bonds in peptides and proteins, *Biochemistry,* 11, 4641, 1972.
22. **Shechter, Y., Patchornik, A., and Burstein, Y.,** Selective chemical cleavage of tryptophanyl peptide bonds by oxidative chlorination with N-chlorosuccinimide, *Biochemistry,* 15, 5071, 1976.

23. **Alexander, N. M.,** Oxidative cleavage of tryptophanyl peptide bonds during chemical- and peroxidase-catalyzed iodinations, *J. Biol. Chem.,* 249, 1946, 1974.

24. **Bornstein, P. and Balian, G.,** Cleavage at Asn-Gly bonds with hydroxylamine, *Meth. Enzymol.,* 47, 132, 1977.

25. **Mahoney, W. C. and Hermodson, M. A.,** High yield cleavage of tryptophanyl peptide bonds by *o*-iodosobenzoic acid, *Biochemistry,* 18, 3810, 1979.

26. **Mahoney, W. C., Smith, P. K., and Hermodson, M. A.,** Fragmentation of proteins with *o*-iodosobenzoic acid: chemical mechanism and identification of *o*-iodoxybenzoic acid as a reactive contaminant that modifies tyrosyl residues, *Biochemistry,* 20, 443, 1981.

Chapter 6

THE MODIFICATION OF CYSTEINE

The sulfhydryl group of cysteine (Figure 1) is in general the most reactive functional group in a protein. Cysteinyl residues are easily alkylated, acylated, arylated, and oxidized (Table 1). The reactivity of cysteine is, as with most other functional groups in proteins, a reflection of the nucleophilic nature of the thiol groups. It is impossible to thoroughly discuss the reactions of protein sulfhydryl groups. The reader is directed to the excellent review by Liu[1] on the properties and reactions of sulfhydryl groups for a more extensive discussion of the chemistry of sulfur in proteins.

Cysteine is far more reactive as the thiolate anion. The pKa for the formation of the thiolate anion is 10.5 with free cysteine but is considerably reduced with the cysteinyl residue in peptide bond. For example, the pKa for the formation of the thiolate anion *N*-acetylcysteine ethyl ester is 8.5 while with *N*-formyl cysteine, it is 9.5. It is useful to compare these values with pKa values for other functional groups as is done in Table 2.

The most useful class of reagents for the modification of cysteinyl residues in proteins has been the α-haloacetates and the corresponding amides. These reagents react with cysteine via a S_N2 reaction mechanism to give the corresponding carboxymethyl or carboxamidomethyl derivatives (see Figure 2). Any of the various α-halo acids or amides can be used. When a rapid reaction is desired, the iodine-containing compounds are used. For example, the reaction of iodoacetate with cysteine is approximately twice as fast as the reaction of bromoacetate and 20 to 100 times as rapid as chloroacetate. There are situations in which fast reaction rates are not necessarily desirable, such as the studies of Gerwin on streptococcal proteinase.[2] This particular study was of considerable importance since it emphasized the importance of microenvironmental effects on the reaction of cysteine with α-haloacids and α-haloamides. Chloroacetic acid was far less effective than chloroacetamide. The sulfhydryl group at the active site of streptococcal proteinase has enhanced reactivity in that modification with iodoacetate readily occurred in the presence of 100- to 1000-fold excess of β-mercaptoethanol or cysteine. The enhanced reactivity of the active site cysteine is also apparent from a comparison of the relative rates of modification of streptococcal proteinase and reduced glutathione. The rate of modification of streptococcal proteinase is 50 to 100 times more rapid than that of glutathione. The unique properties of this cysteine residue can be explained in part by the presence of an adjacent histidyl residue which was demonstrated by an elegant series of studies by Liu.[3] Although histidine residues will react with α-halo acids and amides, the presence of an adjacent cysteine residue precluded the use of this class of reagents to demonstrate the presence of a histidyl residue at the active site of streptococcal proteinase. Liu took advantage of the reversible modification of cysteinyl residues with sodium tetrathionite[4] to modify the active site histidine.

The reaction of chloroacetic acid and chloroacetamide with papain has also yielded interesting results.[5,6] In studies with chloroacetamide, the active site sulfhydryl group of papain reacts at a rate more than tenfold faster than free cysteine (5.78 M^{-1} sec^{-1} vs. 0.429 M^{-1} sec^{-1}).[5] As was the situation with streptococcal proteinase, there are dramatic differences in the rate of reaction of papain with chloroacetic acid and chloroacetamide. Figure 3 shows a first-order and second-order rate plot for the reaction of papain with chloroacetamide. It is of interest to note that the rate of inactivation continues to increase with increasing pH up to approximately pH 10.8 as shown in Figures 4 and 5. Note the difference in the observed behavior at different experimental conditions (here ionic strength was varied). The first-order and second-order rate plots for the reaction of papain with chloroacetic acid as reported by the same investigators are shown in Figure 6. This should be compared with Figure 3. The reaction with chloroacetic acid has a pH optimum of approximately 7 while the optimum

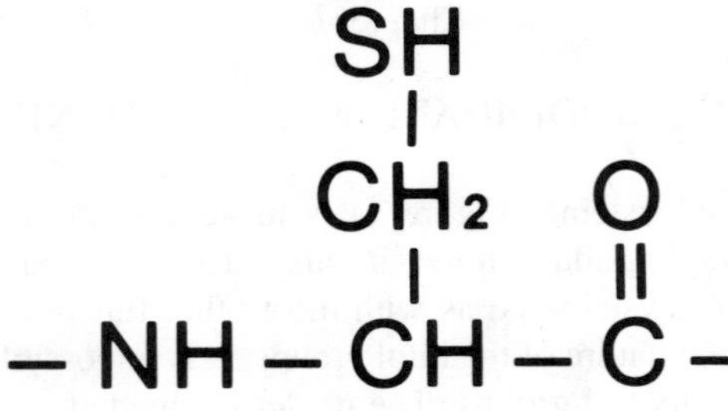

$$SH$$
$$|$$
$$CH_2 \quad O$$
$$| \qquad \|$$
$$-NH-CH-C-$$

FIGURE 1.   The chemical structure of covalently bound cysteine.

Table 1
REACTION OF PROTEIN FUNCTIONAL GROUPS

Amino acid	Alkylation	Acylation	Arylation	Oxidation
Methionine	x	—	x	x
Cysteine	x	x	x	x
Histidine	x	x	x	x
Lysine	x	x	—	—
Tyrosine	x	x	—	x
Tryptophan	—	—	—	x

Table 2
ACID DISSOCIATION VALUES FOR FUNCTIONAL GROUPS IN PROTEINS

Functional Group	pKa
Carboxyl (Asp, Glu)	4.6
Imidazole (His)	7.0
Alpha-amino	7.8
Sulfhydryl (Cys)	8.5
Phenolic Hydroxyl (Tyr)	9.6
Side-chain amino (Lys)	10.5

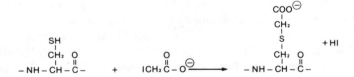

FIGURE 2.   The modification of cysteine with iodoacetic acid to form *S*-carboxymethylcysteine.

for reaction with chloroacetamide is at a pH greater than 9. A comparison of the effect of pH on the reaction of papain with chloroacetic acid and chloroacetamide is shown in Figure 7. This investigation notes the influence of the neighboring histidyl residue as has been discussed for streptococcal proteinase. These data further emphasize the importance of neighboring functional group effects on cysteinyl reactivity in proteins as well as the im-

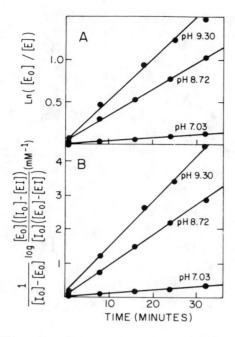

FIGURE 3.   Rate plots for the inactivation of papain by chloroacetamide at various values of pH at low ionic strength (0.07). (From Chaiken, I. M. and Smith, E. L., *J. Biol. Chem.*, 244, 5087, 1969. With permission.)

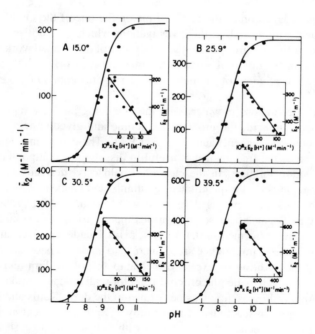

FIGURE 4.   The effect of pH on the second order rate constant for the inactivation of papain by chloroacetamide at high ionic strength (0.50). (From Chaiken, I. M. and Smith, E. L., *J. Biol. Chem.*, 244, 5087, 1969. With permission.)

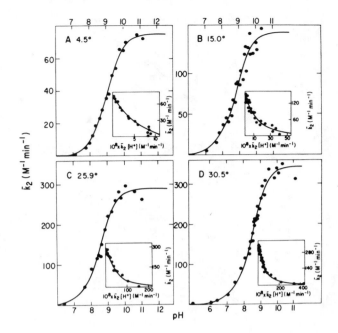

FIGURE 5.   The effect of pH on the second-order rate constant for
the inactivation of papain by chloroacetamide at low ionic strength
(0.07). (From Chaiken, I. M. and Smith, E. L., *J. Biol. Chem.*, 244,
5087, 1969. With permission.)

portance of rigorous evaluation of the effect of pH on the rate of the modification reaction.

A complete consideration of all of the investigations which have utilized α-halo acids for
the modification of sulfhydryl groups is beyond the scope of this limited work. This discussion
will be confined to consideration of some of the most recent work with these derivatives as
well as a consideration of some of the more unique reagents containing the α-halo, β-keto
functions.

The α-halo acids decompose in water, with the rate being far more rapid at alkaline pH.
In the case of iodoacetic acid, the products are iodide and glycolic acid. We recrystallize
the commercially obtained reagents and store over P_2O_5. The compounds are readily soluble
in water. In the case of the free acid, it is useful to dissolve the compound in base prior to
addition to the reaction mixture. In the case of α-haloacetyl derivatives, the resultant *S*-
carboxymethyl cysteine is easily quantitated by amino acid analysis.

Jörnvall and co-workers[7] have used reaction with iodoacetate to probe differences in
structure in wild type β-galactosidase and various mutant forms of the enzyme. The modification reactions were performed in 0.1 *M* Tris, pH 8.1 under nitrogen in the dark. (This
condition is of considerable importance since the α-halo acids are photolabile). The reaction
was terminated by the addition of excess β-mercaptoethanol. Kalimi and Love[8] have examined the reaction of the hepatic glucorticoid-receptor with iodoacetamide in 0.010 *M* Tris-
0.25 *M* sucrose. Again, this reaction was performed in the dark. Kallis and Holmgren[9] have
examined the differences in reactivity of two sulfhydryl groups present at the active site of
thioredoxin. The pH dependence of the reaction with iodoacetate suggested that one group
had a pKa value of 6.7 while the second was 9.0. Iodoacetamide showed the same pH
dependence but the rate of reaction was approximately 20-fold greater than with iodoacetate.
For example, at pH 7.2, the second order rate constant for reaction with iodoacetate was
5.2 M^{-1} sec^{-1} while it was 107.8 M^{-1} sec^{-1} for iodoacetamide. The results from this study
are shown in Figure 8. The low pK of one of the sulfhydryl groups was suggested to be a

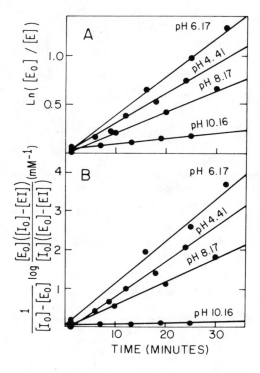

FIGURE 6. Rate plots for the inactivation of papain by chloroacetic acid at various values of pH at low ionic strength (0.07). (From Chaiken, I. M. and Smith, E. L., *J. Biol. Chem.*, 244, 5095, 1969. With permission.)

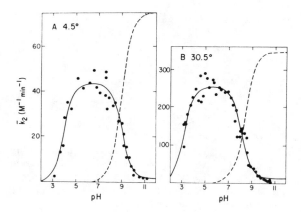

FIGURE 7. The effect of pH on the second-order rate constant for the inactivation of papain by chloroacetic acid at low ionic strength (0.07). The broken lines are the theoretical curves for the reaction of papain with chloroacetamide under the same reaction conditions. (From Chaiken, I. M. and Smith, E. L., *J. Biol. Chem.*, 244, 5095, 1969. With permission.)

reflection of the presence of an adjacent lysine residue. Mikami and co-workers have examined the inactivation of soybean β-amylase with iodoacetamide and iodoacetate.[10] Inactivation with iodoacetamide occurred approximately 60 times more rapidly than with

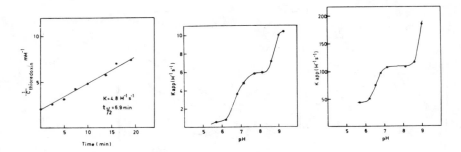

FIGURE 8.   The left figure shows a time course for the reaction of thioredoxin and iodoacetic acid at pH 7.2. Analysis of this data yields a single second-order rate constant of 4.8 M^{-1} sec^{-1} and a halftime of 6.9 min. The center figure shows the effect of pH on the second-order rate constant for the reaction between iodoacetic acid and thioredoxin. The figure on the right shows the effect of pH on the second-order rate constant for the reaction between iodoacetamide and thioredoxin. (From Kallis, G.-B. and Holmgren, A., *J. Biol. Chem.*, 255, 10261, 1980. With permission.)

ANALYSIS OF DISULFIDE BONDS

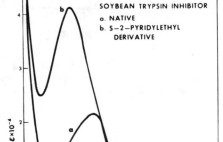

FIGURE 9.   The UV absorption spectra of soybean trypsin inhibitor and the *S*-pyridylethylcysteinyl derivative of soybean trypsin inhibitor at a concentration of 0.5 mg/mℓ (23 μM) at pH 3.0 in 0.05 *M* glycine-HCl. (From Friedman, M., Krull, R. H., and Cavins, J. F., *J. Biol. Chem.*, 245, 3868, 1970. With permission.)

iodoacetate at pH 8.6. Hempel and Pietruszko[11] have shown that human liver alcohol dehydrogenase is inactivated by iodoacetamide but not by iodoacetic acid. These experiments were performed in 0.030 *M* sodium phosphate, pH 7.0 containing 0.001 *M* EDTA.

It should be noted that the reaction of sulfhydryl groups with iodoacetate[12] is still extensively used in the preparation of proteins for primary structure analysis, although pyridylethylation[13,14] is proving to be quite useful (Figure 9).

A related compound which has proven useful is bromotrifluoroacetone.[15-17] This derivative can be used to introduce ^{19}F for nuclear magnetic resonance studies of the microenvironment surrounding the site of modification. This reaction is shown in Figure 10.

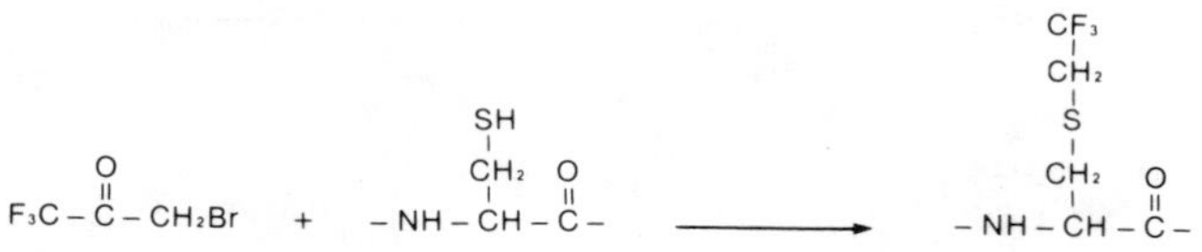

$$F_3C-\overset{\overset{\displaystyle O}{\|}}{C}-CH_2Br \quad + \quad -NH-\overset{\overset{\displaystyle CH_2}{|}\;\underset{\displaystyle SH}{}}{CH}-\overset{\overset{\displaystyle O}{\|}}{C}- \quad \longrightarrow \quad -NH-\overset{\overset{\displaystyle CH_2}{|}}{CH}-\overset{\overset{\displaystyle O}{\|}}{C}-$$

FIGURE 10.   The reaction of bromotrifluoroacetone with cysteine.

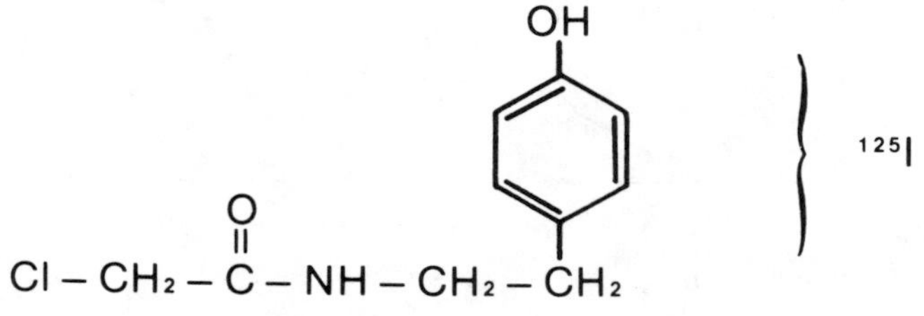

$$CI-CH_2-\overset{\overset{\displaystyle O}{\|}}{C}-NH-CH_2-CH_2 \qquad \Big\}\; {}^{125}I$$

FIGURE 11.   The structure of *N*-chloracetyl-[^{125}I]-iodo-tyramine.

$$\text{imidazole}-CH_2-\overset{\overset{\displaystyle Br}{|}}{CH}-COOH$$

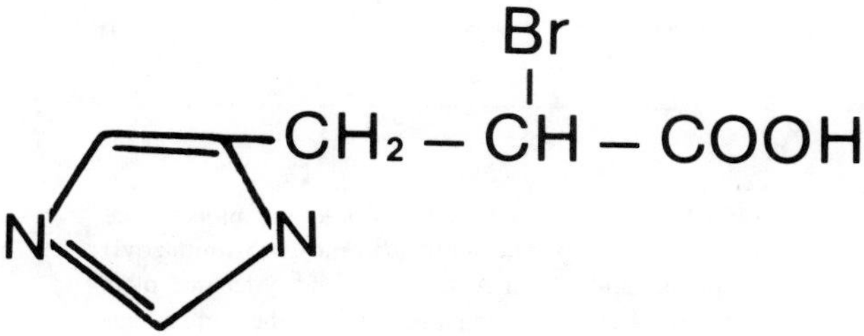

FIGURE 12.   The structure of α-bromo-β-(5-imidazoyl) propionic acid.

Dahl and McKinley-McKee[18] have made a rather detailed study of the reaction of alkyl halides with thiols. It is emphasized that reactivity of alkyl halides not only depends on the halogen but also on the nature of the alkyl groups.

An alpha-halo compound related to chloroacetic acid, *N*-chloroacetyl-[^{125}I]-iodo-tyramine has been developed for the introduction of a radiolabel with high specific activity into proteins containing sulfhydryl groups[19] (see Figure 11).

The synthesis of α-bromo-β-(5-imidazoyl) propionic acid (Figure 12) has been reported by Yankeelov and Jolley.[20] This compound is stable in aqueous solution under neutral conditions and only slight decomposition is seen at pH 10.0. At pH 8.0, the rate of reaction with free cysteine was 0.0036 M^{-1} sec^{-1}. The reaction of this compound with papain has been explored by the same investigators.[21] Figure 13 shows first- and second-order rate plots for the reaction of DL-α-bromo-β-(5-imidazoyl) propionate with papain. The pH dependence for this reaction is presented in Figure 14. It is noted that the carboxyl group of the reagent is of critical importance in this reaction as neither 5-vinylimidazole or 5-(2-bromoethyl) imidazole reacts with papain under these reaction conditions. The rate of the modification of the active site sulfhydryl group is approximately 100 times faster than the reaction with free cysteine.

Chloroacetol phosphate (Figure 15) is a compound related to the α-haloacids described above. The reaction of this compound with rabbit muscle aldolase has been studied.[22,23] This compound was developed as a potential affinity label for aldolase, triosephosphate isomerase, and glycerophosphate dehydrogenase.[24]

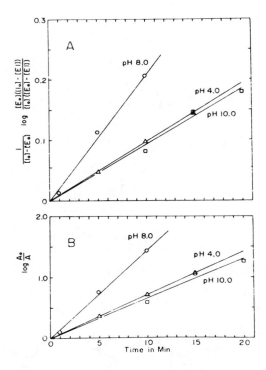

FIGURE 13.   Second- and first-order rate plots for the
inactivation of papain with α-bromo-β-(5-imidazoyl)
propionic acid. Panel A shows second-order rate plots
at three pH values while panel B shows the corresponding first-order plots for the same reactions. (From Jolley,
C. J. and Yankeelov, J. A., Jr., *Biochemistry,* 11, 164,
1979. With permission.)

While it is likely that compounds such as iodoacetate, chloroacetate and their derivatives
have been the most popular class of compounds for the modification of sulfhydryl groups
over the past 2 decades, the use of *N*-ethylmaleimide and its derivatives has also been of
considerable value.

N-ethylmaleimide reacts with sulfhydryl groups (Figure 16) in proteins with considerable
specificity.[25-27] The effect of pH on the reaction of free cysteine and *N*-ethylmaleimide is
shown in Figure 17. These determinations were performed in the pH range of 3 to 5 and
these investigators estimated that a rate constant of $1.53 \times 10^3 \, M^{-1} \, sec^{-1}$ would be obtained
at pH 7.0. This reaction can be followed spectrophotometrically by the decrease in absorbance
at 300 nm, the absorbance maximum of *N*-ethylmaleimide. This reaction product yields *S*-
succinyl cysteine on acid hydrolysis. Although the reagent is reasonably specific for cysteine,
reaction with other nucleophiles must be considered.[28] The use of reaction with *N*-ethyl-
maleimide in the identification of selenocysteine has been proposed.[29] A "diagonal" procedure for the isolation of cysteine-containing peptides modified with *N*-ethylmaleimide has
been reported (Figure 18).[30] This procedure is based on the hydrolysis of the reaction product
of cysteine and *N*-ethylmaleimide to cysteine-*S*-*N*-ethyl succinamic acid generating a new
negative charge. The extinction coefficient of *N*-ethylmaleimide is $620 \, M^{-1} \, cm^{-1}$ at 302
nm.[25]

There has been a considerable amount of interest recently in maleimide derivatives. For
example, various derivatives of maleimide provide the basis for the design of cross-linking
reagents (see Volume II, Chapter 5). Brown and Matthews[31,32] have studied the reaction of

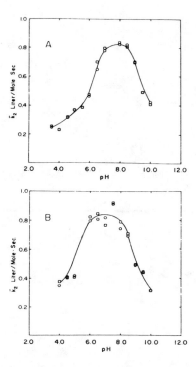

FIGURE 14. The effect of pH on the inactivation of papain by α-bromo-β-(5-imidazoyl) propionic acid(BIP) at 37°C. Panel A describes a study performed with papain activated with β-mercaptoethanol-EDTA while panel B describes a study performed with papain activated with dithiothreitol-EDTA. (From Jolley, C. J. and Yankeelov, J. A., Jr., *Biochemistry,* 11, 164, 1972. With permission.)

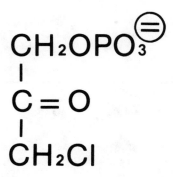

FIGURE 15. The structure of chloroacetyl phosphate.

lactose repressor protein with *N*-ethylmaleimide, two spin-label derivatives of *N*-ethylmaleimides and a fluorophore derivative. The structures of these various derivatives are shown in Figure 19. The investigators demonstrated that three sulfhydryl residues present in *Escherichia coli* lactose repressor protein monomer had distinctly different reaction characteristics. These data are presented in Figure 20. In these studies the protein was modified with *N*-ethylmaleimide or one of the derivatives shown in Figure 19 (the modification reactions were performed in 0.24 *M* potassium phosphate, pH 7.0 containing 5% (v/v) glycerol under nitrogen for 1 to 4 hr). The extent of reaction was determined by the reaction of the remaining

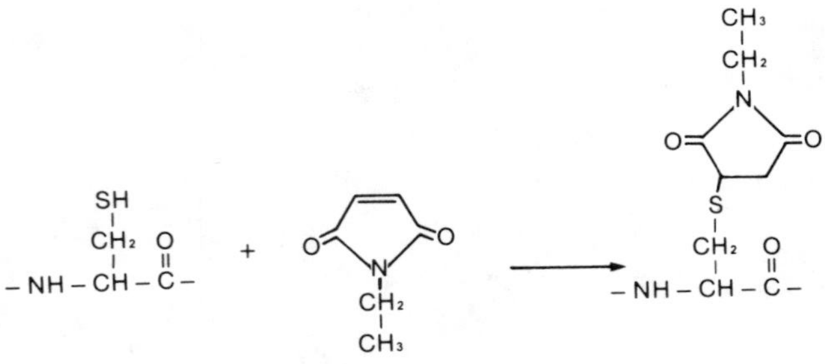

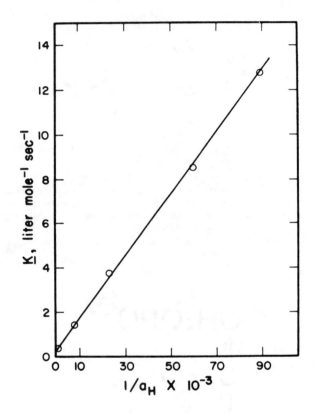

FIGURE 16.   The reaction of *N*-ethylmaleimide with cysteine.

FIGURE 17.   The pH dependence of the reaction of *N*-ethyl-
maleimide with cysteine. (From Gorin, G., Martic, P. A., and
Doughty, G., *Arch. Biochem. Biophys.*, 115, 593, 1966. With
permission.)

free sulfhydryl groups with 2-chloromercuri-4-nitrophenol. Modification at a specific cysteinyl residue was determined by reaction of the modified protein with 2-bromoacetamido-
4-nitrophenol and quantitation of the three 5-(2-acetamido-4-nitrophenol)-cysteine-contain-
ing peptides following enzymatic digestion of the modified protein and gel filtration on G-
50 Sephadex (0.1 *M* NH$_4$HCO$_3$ containing 2% (w/v) sodium dodecyl sulfate). The spin-
labeled compounds showed the same pattern of reaction with the three cysteinyl residues as
seen with *N*-ethylmaleimide. The fluorophore-derivative (*N*-(3-pyrene)maleimide) shows a
slightly different reaction pattern. Le-Quoc and colleagues have examined the effect of the

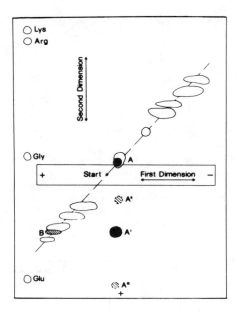

FIGURE 18. A "diagonal" method for the identification of cysteine peptides. A mixture of chymotryptic peptides containing a peptide with a cysteinyl residue alklylated with radiolabeled *N*-ethylmaleimide was subjected to diagonal electrophoresis with intervening exposure to ammonia for 5.5 hr at 35°C. The treatment with ammonia results in the hydrolysis of the *N*-ethylsuccinimide to *N*-ethylsuccinamic acid thus generating a new negative charge permitting the resolution of the labeled peptides from the unlabeled peptides during the second electrophoresis. (From Gehring, H. and Christen, P., *Analyt. Biochem.*, 107, 358, 1980. With permission.)

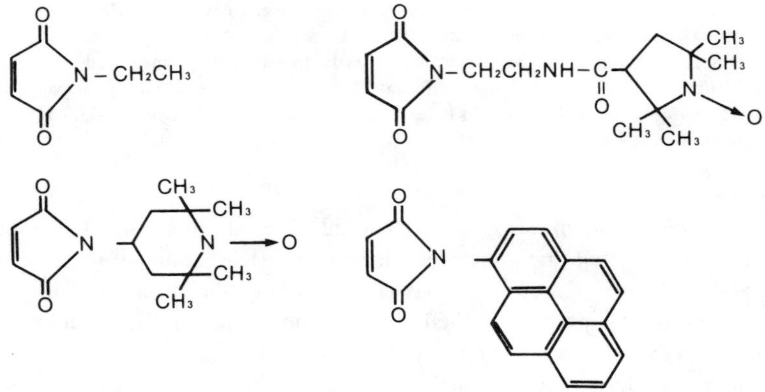

FIGURE 19. The structure of *N*-ethylmaleimide and some derivatives. The structure in the upper left is *N*-ethylmaleimide, the lower left shows 4-maleimido-2,2,6,6,-tetramethylpiperidinoxyl, the upper right shows 3-(2-maleimidoethyl)carbamoyl-2,2,5,5-tetramethyl-1-pyrrolidinyloxyl and the lower right shows *N*-(3-pyrene)maleimide.

nature of the *N*-substituent groups on the rate of sulfhydryl group modification in succinate dehydrogenase.[33] The derivatives used were *N*-ethylmaleimide, *N*-butylmaleimide and *N*-benzylmaleimide. The most reactive thiol groups in succinate dehydrogenase are probably located in an apolar environment since the benzyl derivative reacted twice as fast as the

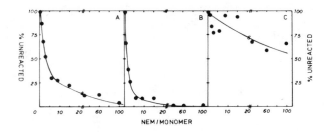

FIGURE 20.   The reaction of *lac* repressor protein with *N*-ethyl-maleimide at specific cysteine residues as function of *N*-ethylmaleimide concentration. Panel A shows reaction at cysteine-107, panel B shows reaction at cysteine-140, while panel C shows reaction at cysteine-281. (From Brown, R. D. and Matthews, K. S., *J. Biol. Chem.*, 254, 5128, 1979. With permission.)

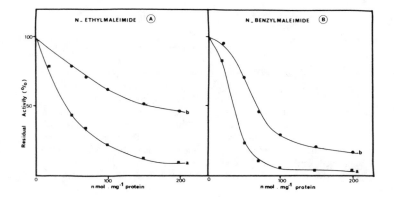

FIGURE 21.   The inactivation of succinate dehydrogenase with *N*-ethylmaleimide (panel A) or *N*-benzylmaleimide (panel B) as a function of reagent concentration. Reaction a was performed with enzyme preparations preincubated with 50 m*M* succinate, 3 m*M* thenoyltrifluoroacetone and 10 μg rotenone. Reaction was performed with no additions other than the maleimide derivatives, in 0.05 *M* sodium phosphate, pH 7.6. (From Le-Quoc, K., Le-Quoc, D., and Guademer, Y., *Biochemistry*, 20, 1705, 1981. With permission.)

ethyl derivative. This is shown in Figures 21 and 22. The rate plots shown in Figure 22 are biphasic in nature suggesting that there are at least two different sites of modification which influence catalytic activity. Careful analysis (Figure 23) of the reaction of the membrane-bound succinate dehydrogenase with *N*-ethylmaleimide (these modification reactions were performed in 0.050 *M* sodium phosphate, pH 7.6, at 37°C) shows this to be the situation.

Another maleimide derivative which is of interest is *N*-(7-dimethylamino-4-methylcou-marinyl) maleimide.[34-36]

The conversion of cysteinyl residues to the *S*-cyanyl derivatives has also received considerable attention as the modification can be accomplished with a chromogenic reagent such as 2-nitro-5-thiocyanobenzoic acid (Figure 24).

Although the formation of the *S*-cyano derivative (Figure 25) is the predominant reaction, Degani and Degani[37] have also demonstrated formation of the mixed disulfide with mercaptonitrobenzoate as well. These investigators studied the reaction of rabbit muscle creatine kinase with 2-nitro-5-thiocyanobenzoic acid in 0.02 *M* Tris, pH 7.8 containing 0.25 m*M* EDTA with a 2.5- to 10-fold molar excess of reagent. The expected reaction was that shown

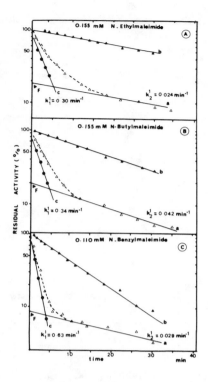

FIGURE 22.   Comparison of the rate of inactivation of membrane-bound succinate dehydrogenase by three different *N*-maleimide derivatives. Panel A utilized 0.155 m*M* *N*-ethylmaleimide, panel B utilized 0.155 m*M* *N*-butylmaleimide, while panel C utilized 0.110 m*M* *N*-benzylmaleimide. The reactions were performed at 25°C in 0.05 *M* sodium phosphate, pH 7.6. In reaction A, the enzyme preparation was preincubated with 50 m*M* β-hydroxybutyrate while in reaction B, the enzyme preparation was preincubated with 50 m*M* succinate, 3 m*M* thenoyltrifluoroacetone, and 10 μg rotenone. (From Le-Quoc, K., Le-Quoc, D., and Guademer, Y., *Biochemistry*, 20, 1705, 1981. With permission.)

in Figure 25. While reaction of creatine kinase with 5,5′-dithiobis (2-nitrobenzoic acid) resulted in the modification of two sulfhydryl groups with greater than 99% loss of activity, reaction with 2-nitro-5-thiocyanobenzoic acid resulted in an equivalent loss of activity with apparent modification only at a single sulfhydryl residue as judged by the release of 2-mercapto-5-nitrobenzoic acid. There were not, however, any free sulfhydryl groups remaining after the reaction with 2-nitro-5-thiocyanobenzoic acid. Spectral analysis of the modified enzyme (Figure 26) was consistent with the incorporation of 1 mol of reagent per mol of protein ($\epsilon_{330} = 7500$) as would result from the reaction shown in Figure 27. The denatured enzyme showed only the *S*-cyanylation reaction. Reaction of the modified protein with cyanate (0.11 *M* potassium cyanate, pH 9.5) resulted in the conversion to the cyanylated derivative as shown in Figure 28. This reaction was associated with the return of 75% of the enzymatic activity of the native enzyme. Reaction of creatine with kinase with 5,5′-dithiobis (2-nitrobenzoic acid) resulted in the modification of approximately 2 sulfhydryl groups (1.75 to 1.8 mol/mol of enzyme) with an almost complete loss of enzyme activity (greater than 99.5% loss of activity). The results of the cyanolysis of this modified enzyme are shown in Figure 29. It is readily apparent that there are two phases for the release of 2-mercapto-5-nitrobenzoic acid. The conditions for this reaction are the same as those used for Figure 28. The more rapid initial reaction appeared to be related to the return of enzymatic

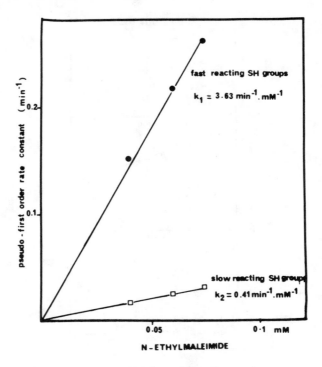

FIGURE 23.   The influence of *N*-ethylmaleimide concentration on the pseudo first-order rate constants for the inactivation of succinate dehydrogenase. The first-order rate constants were obtained from graphs similar to those shown in Figure 22 utilizing varying concentrations of *N*-ethylmaleimide. The second-order rate constants were determined from the slopes of the lines. (From Le-Quoc, K., Le-Quoc, D., and Gaudemer, Y., *Biochemistry*, 20, 1705, 1981. With permission.)

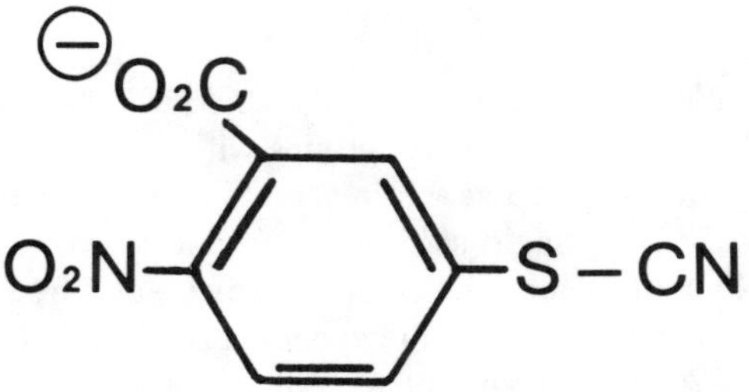

FIGURE 24.   The structure of 2-nitro-5-thiocyano-benzoic acid.

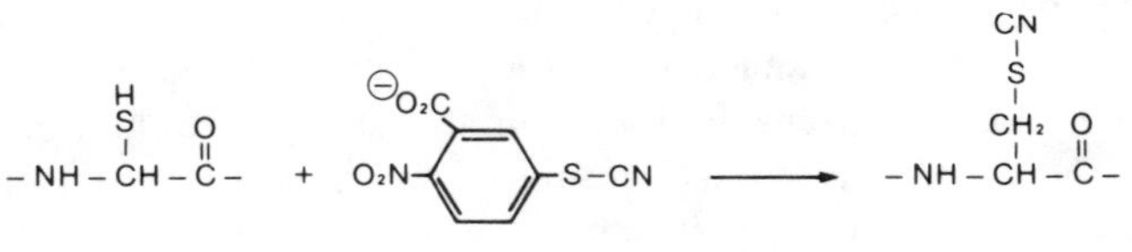

FIGURE 25.   The reaction of 2-nitro-5-thiocyanobenzoic acid with sulfhydryl groups in proteins.

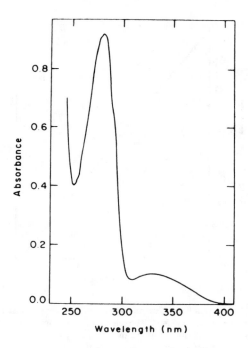

FIGURE 26. The UV absorption spectra of rabbit muscle creatine kinase after modification with 2-nitro-5-thiocyanobenzoic acid (NTCB). The spectra were obtained in 0.02 *M* Tris-acetate, pH 7.0. (From Degani, Y. and Degani, C., *Biochemistry,* 18, 5917, 1979. With permission.)

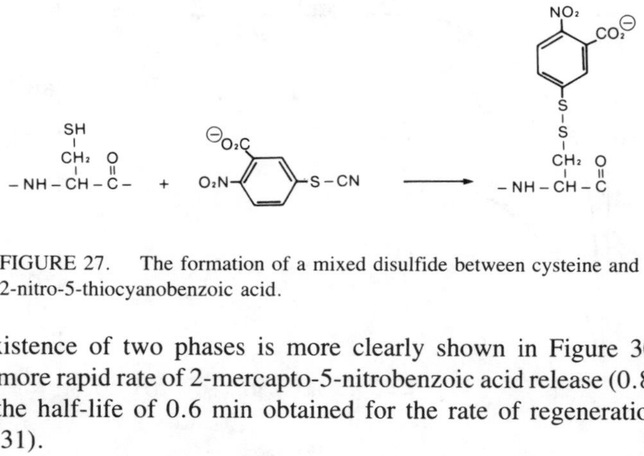

FIGURE 27. The formation of a mixed disulfide between cysteine and 2-nitro-5-thiocyanobenzoic acid.

activity. The existence of two phases is more clearly shown in Figure 30. The half-life obtained for the more rapid rate of 2-mercapto-5-nitrobenzoic acid release (0.8 min) compares favorably with the half-life of 0.6 min obtained for the rate of regeneration of enzymatic activity (Figure 31).

The formation of 2-mercapto-5-nitrobenzoic acid, which occurs with the reaction of 2-nitrothiocyanobenzoic acid with thiols to form *S*-cyano derivatives, can be used for the quantitative determination of sulfhydryl groups. 2-Mercapto-5-nitrobenzoic acid has an absorbance maximum at 412 nm with a molar extinction coefficient of 13,600 M^{-1} cm^{-1}.[38] Pecci and co-workers[39] have characterized the reaction of rhodanese with 2-nitrothiocyanobenzoic acid. These investigators used a 1.3 molar excess of reagent in 0.050 *M* phosphate buffer, pH 8.0 at 18°C. The reaction was followed spectrophotometrically by the release of 2-mercapto-5-nitrobenzoic acid and was complete after 6 hr.

Cleavage at *S*-cyano residues has been reported by Marshall and Cohen.[40] In these studies

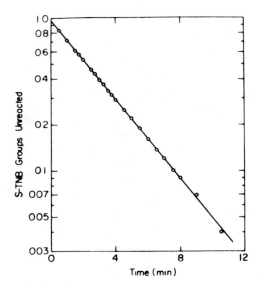

FIGURE 28.   Cyanolysis of rabbit muscle creatine kinase modified with 2-nitro-5-thiocyanobenzoic acid. The release of 2-mercapto-5-nitrobenzoic acid is shown upon treatment of the modified enzyme with 0.11 M [^{14}C] KCN in 0.125 M potassium phosphate, pH 9.5. (From Degani, Y. and Degani, C., *Biochemistry*, 18, 5917, 1979.

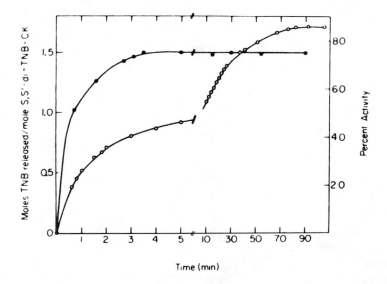

FIGURE 29.   Cyanolysis of rabbit muscle creatine kinase modified with 5,5'-dithiobis(2-nitrobenzoic acid) (S,S'-di-TNB-CK). The modified enzyme was reacted with 0.11 M KCN in 0.125 M potassium phosphate, pH 9.5. The reaction was followed by the release of 2-mercapto-5-nitrobenzoic acid (○) and by the regeneration of catalytic activity (●). (From Degani, Y. and Degani, C., *Biochemistry*, 18, 5917, 1979. With permission.)

the enzyme was first reacted with 5,5'-dithiobis (2-nitrobenzoic acid) in 0.020 M 4-morpholinepropanesulfonic acid — 0.1 M KCl, pH 7.1 for 3 hr at 25°C. Conversion to the 5-cyano derivative was accomplished by reaction in 0.2 M KCl, pH 8.1. Cleavage was

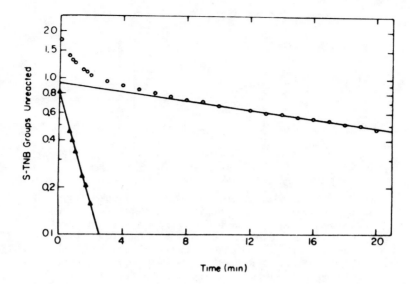

FIGURE 30.   Time course plot (semi-log) for the release of 2-mercapto-5-nitro-benzoic acid from *S,S'*-di-TNB-CK under the conditions described in the legend for Figure 29. The open circles represent the experimental data points while the closed triangles are values obtained by subtracting the contribution of the slow reaction from the observed data. (From Degani, Y. and Degani, C., *Biochemistry*, 18, 5917, 1979. With permission.)

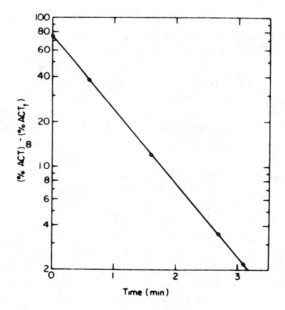

FIGURE 31.   A semi-log plot for the rate of recovery of catalytic activity during the cyanolysis of *S,S'*-di-TNB-CK as described in Figure 29. (From Degani, Y. and Degani, C., *Biochemistry*, 18, 5917, 1979. With permission.)

accomplished by incubation at pH 8.0 at 50°C for 24 hr. The reaction of 2-nitro-5-thiocy-anobenzoic acid with phosphofructokinase has been studied by Ogilvie.[41] Approximately 1

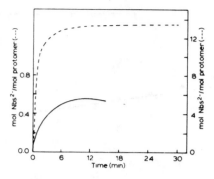

FIGURE 32.   The release of 5-mercapto-2-nitrobenzoate dianion (Nbs²⁻)
during the sequential reaction of phosphofructokinase with 2-nitro-5-thio-
cyanobenzoic acid and 5,5'-dithiobis(2-nitrobenzoic acid) as determined
by the increase in absorbance at 412 nm. The solid line (left ordinate)
represents the results of the reaction of 2-nitro-5-thiobenzoic acid with
phosphofructokinase (1.06 mol/mol protomer) at pH 7.2 resulting in the
cyanylation of 0.57 mol/mol of cysteine per mol of protomer. The dashed
line (right ordinate) represents the results of the reaction of 5,5'-dithiobis(2-
nitrobenzoic acid) (60-fold molar excess with respect to protomer) in 2%
sodium dodecyl sulfate(SDS), pH 7.2 with phosphofructokinase containing
0.57 mol of cyanylated cysteine per mole of protomer. (From Ogilvie, J.
W., *Biochim. Biophys. Acta*, 622, 277, 1980. With permission.)

mol of cysteine is available for modification in the native enzyme with an approximately
stoichiometric excess of reagent (1.06 mol 2-nitro-5-thiocyanobenzoic acid per mole enzyme
protomer)(Figure 32). A similar extent of modification is observed on reaction of the native
enzyme with 5,5'-dithiobis(2-nitrobenzoic acid). The modification with 2-nitro-5-thiocy-
anobenzoic acid was performed in 0.025 M glycylglycine — 0.025 M sodium phosphate,
pH 7.2 (containing 1 mM EDTA, 0.4 mM fructose-6-phosphate, and 0.1 mM ATP) at 24°C.
The rate and extent of modification was determined by following the rate of 2-nitro-5-
thiobenzoic acid release at 412 nm. Cleavage at the S-cyanocysteinyl residue is accomplished
by incubation of the modified protein in 0.2 M Tris-acetate, pH 8.1 containing 2% sodium
dodecylsulfate at 37°C. Approximately 20% of the total phosphofructokinase was cleaved
after 48 hr of incubation at 37°C. This would correspond to approximately 40% cleavage
of the S-cyanylated protein.

Kindman and Jencks[42] have reported interesting observations on the reaction of 2-nitro-
thiocyanobenzoic acid with succinyl-CoA:3 ketoacid coenzyme A transferase. These investigators conclude that the thiol group which reacts with this reagent to form the S-cyano
derivative was not essential to activity. Reaction of the enzyme with 2-nitrothiocyanobenzoic
acid was performed in 0.2 M potassium borate, pH 8.1 (Tris buffers as well as other buffers
with nucleophilic characteristics should be avoided because of reaction with the reagent).
While the reaction to yield the S-cyano derivative had no effect on catalytic activity, the
formation of a mixed disulfide with 2-mercapto-5-nitrobenzoic acid via either 5,5'-dithiobis-
(2-nitrobenzoic acid) or 2-nitrothiocyanobenzoic acid (2-nitro-5-(thiocyanato) benzoic acid)
resulted in an inactive enzyme (Figure 33).

One of the most popular reagents for the modification and determination of sulfhydryl
group has evolved from the early studies of Ellman[38] on 5,5-dithiobis-(2-nitrobenzoic) acid
(Figure 34). Reaction with sulfhydryl groups in proteins results in the release of 2-nitro-5-
mercaptobenzoic acid (Figure 35), which has a molar extinction coefficient of 13,600 M^{-1}
cm^{-1} at 410 nm. Recent examples of the use of this reagent have included studies on *E.
coli* citrate synthase[43] and D-amino acid transaminase[44] (0.1 M Tris, 0.002 M EDTA, pH

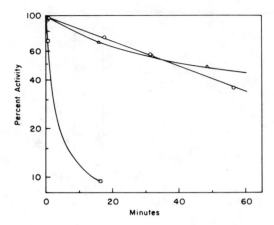

FIGURE 33. 5,5'-dithiobis(2-nitrobenzoic acid)(DTNB) inactivation of succinyl-CoA:3-ketoacid coenzyme A transferase previously modified with 2-nitro-5-(thiocyanato)benzoic acid (NTCB)($\triangle$), NTCB-modified enzyme separated from reagent by gel filtration ($\bigcirc$), and NTCB-modified enzyme preincubated with 0.095 *M* dithiothreitol for 1 hr in 0.2 *M* Tris-sulfate, pH 8.1 prior to gel filtration ($\square$). (From Kindman, L. A. and Jencks, W. P., *Biochemistry*, 20, 5183, 1981. With permission.)

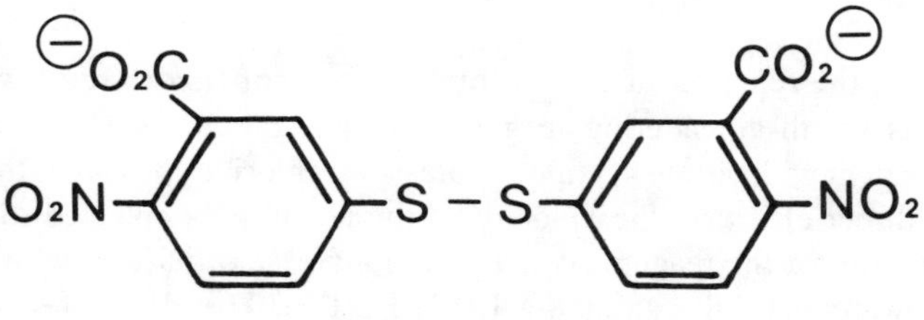

FIGURE 34. The structure of 5,5-dithio*bis*(2-nitrobenzoic acid).

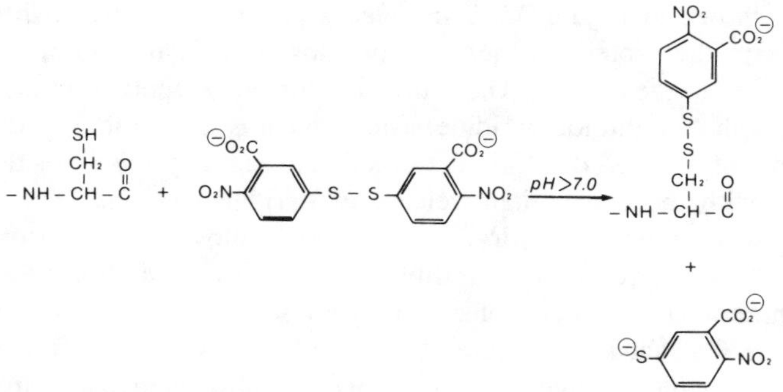

FIGURE 35. The reaction of 5,5-dithio*bis*(2-nitrobenzoic acid) with cysteinyl residues in proteins.

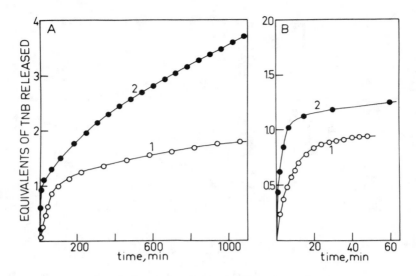

FIGURE 36.   The reaction of *Escherichia coli* citrate synthase with 5,5'-dithiobis(2-nitrobenzoic acid)(DTNB). The reaction was performed at pH 7.8. In panel A, KCl was not present; experiment 1 contained 50μM DTNB and 13.4 μM enzyme and experiment 2 contained 1500 μM DTNB and 132 μM enzyme. In panel B, 0.1 *M* KCl was present; experiment 1 contained 100 μM DTNB and 30.7 μM enzyme and experiment 2 contained 1100 μM DTNB and 293 μM enzyme. The extent of reaction was monitored by the release of the thiophenolate anion of 5-thio-2-nitrobenzoic acid (TNB). (From Talgoy, M. M., Bell, A. W., and Duckworth, H. W., *Can. J. Biochem.*, 57, 822, 1979. With permission.)

7.5). The study on the reaction of 5,5'-dithiobis-(2-nitrobenzoic acid) with the bacterial citrate synthase is worth considering in greater detail. Figure 36 shows the time for the reaction of this reagent with citrate synthase. Potassium chloride stimulates the rate of reaction apparently by a direct effect on the velocity of the reaction as opposed to a change in the affinity of the protein for the reagent. A maximum increase of 85-fold with KCl is observed in 0.02 *M* Tris buffer, pH 7.8 (containing 1.0 *M* EDTA). The effect of salt is not a general effect on the reactivity of sulfhydryl groups since 0.1 *M* KCl decreases the rate of reaction of 5,5'-dithiobis-(2-nitrobenzoic acid) with coenzyme A. It is of interest that there is the release of 5-thio-2-nitrobenzoate from the modified enzyme after removal of reagents by gel filtration as shown in Figure 37. This release presumably reflects the formation of a cystine disulfide in the protein as there are two fewer sulfhydryl groups in the modified protein as compared to the control. These investigators also reported on the modification of citrate synthase with 4,4'-dithiodipyridine. This reagent is similar to 5,5'-dithiobis-(5-nitro-benzoic acid) in that a mixed disulfide is formed between a cysteinyl residue in the protein and the reagent with the concomitant release of pyridine-4-thione. The reaction of 4,4-dithiodipyridine with protein sulfhydryl groups can be followed by spectroscopy ($\epsilon_{324 \text{ nm}} = 19,800 \ M^{-1} \ cm^{-1}$). The reaction is readily reversed by the addition of a reducing agent such as dithiothreitol. The reaction of citrate synthase and 4,4-dithiodipyridine in 0.02 *M* Tris, pH 7.8 at 21°C is shown in Figure 38. Figure 39 shows the effect of prior reaction of citrate synthase with one of the above reagents on subsequent reactivity with the other reagent. Modification of one sulfhydryl group with either reagent greatly reduced both the rate and extent of subsequent reaction with the other reagent. The reaction of 5,5'-dithiobis(2-nitrobenzoic acid) with D-amino acid transaminase also provides an illustration of the use of this reagent.[44] These studies were performed in 0.1 *M* Tris, pH 7.5; the results are shown in Figure 40. An extinction coefficient of 14,140 $M^{-1}cm^{-1}$ for the 2-nitro-5-thiobenzoate

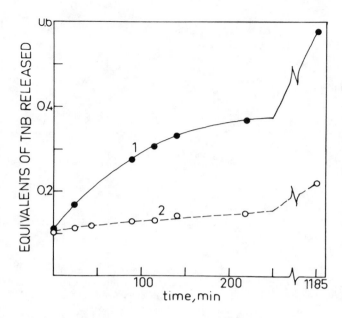

FIGURE 37.   The spontaneous release of TNB from DTNB-modified citrate synthase after the removal of reagents by gel filtration. The solvent was 0.02 *M* Tris-HCl, pH 7.8—1. m*M* EDTA. In experiment 2 the solvent also contained 0.1 *M* KCl. The release of TNB was monitored by the increase in absorbance at 412 nm. (From Talgoy, M. M., Bell, A. W., and Duckworth, H. W., *Can. J. Biochem.*, 57, 822, 1979. With permission.)

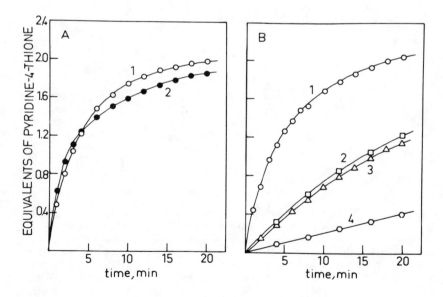

FIGURE 38.   The reaction of citrate synthase with 4,4'-dithiodipyridine (4,4'-PDS). The reactions were performed at pH 7.8 in 0.02 *M* Tris-HCL, 1 m*M* EDTA. In panel A, KCl was absent in curve 1 and present at a concentration of 0.1 *M* in curve 2. In panel B, curve 1 contains only the Tris-EDTA buffer, curve 2 contains 1.66 m*M* 5'-AMP, curve 3 contains 32 μ*M* NADH, and curve 4 contains 2.94 m*M* ADP-ribose. The enzyme concentration was 21.8 μ*M* in all experiments while the concentration of 4,4'-PDS was 50 μ*M*. (From Talgoy, M. M., Bell, A. W., and Duckworth,. W., *Can J. Biochem.*, 57, 822, 1979. With permission.)

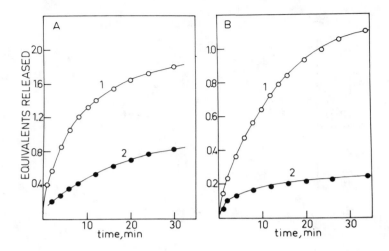

FIGURE 39.   The effect of the prior modification of citrate synthase with one sulfhydryl reagent upon subsequent reaction with a different reagent. In panel A curve 1 represents the time course of the modification of citrate synthase with 4,4′-PDS; curve 2 represents a preparation of citrate synthase which was first modified with DTNB in the presence of 0.1 M KCl, subjected to gel filtration and allowed to react with 4,4′-PDS as in curve 1. At 32 min the difference between curve 1 and curve 2 is 0.99 groups per subunit. In panel B, curve 1 represents the time course for the reaction of citrate synthase with DTNB in the presence of 0.1 M KCl; curve 2 represents the time course of reaction of a preparation of citrate synthase previously modified with 4,4′-PDS, subjected to gel filtration and then allowed to react with DTNB under the same conditions as curve 1. At 34 min, the difference between curves 1 and 2 is 0.86 groups per subunit. (From Talgoy, M. M., Bell, A. W., and Duckworth, H. W., *Can. J. Biochem.*, 57, 822, 1979. With permission.)

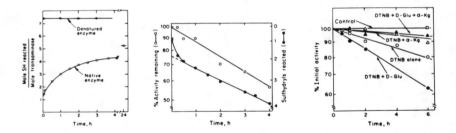

FIGURE 40.   The left figure shows the time course for the modification of the sulfhydryl groups of D-amino acid transaminase with DTNB in 0.1 M Tris, pH 7.2, 2.0 mM EDTA. The absorption at 412 nm was followed as function of time and the number of sulfhydryl groups modified was determined using an extinction coefficient of 14,140. The experiment with the denatured enzyme was performed in the presence of 6.4 M guanidine hydrochloride. The center figure represents the temporal correlation between the extent of sulfhydryl group modification and loss of D-amino acid transaminase activity. The right hand figure shows the effect of substrates on the inactivation of D-amino acid transaminase by DTNB. (From Soper, T. S., Jones, W. M., and Manning, J. M., *J. Biol. Chem.*, 245, 10901, 1979. With permission.)

anion was used in these studies. Approximately one half of the sulfhydryl groups are available for reaction with this reagent in the native enzyme. Reaction with 5,5′-dithiobis-(2-nitro-benzoic acid) does result in the loss of activity and this loss of activity appears to be correlated with the modification of one of the more slowly reacting cysteinyl residues. Other proteins

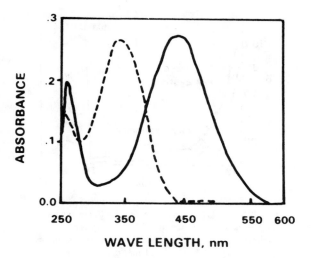

FIGURE 41. The UV absorption spectrum of 6-selenobis(3-nitrobenzoic acid)(SNB^{2-})(25.9 μM) and 6,6-diselenobis(3-nitrobenzoic acid)(DSNB)(12.9 μM) in 0.2 M Tris-HCl, pH 8.2, 1 mM EDTA. (From Luthra, M. P., Dunlap, R. B., and Odom, J. D., *Analyt. Biochem.*, 117, 94, 1981. With permission.)

which have recently been studied with this reagent include rat brain nicotinic-like acetylcholine receptors[45] (calcium-containing Ringers solution, pH 7.4), lipophilin from human myelin[46] (0.001 M glycylglycine — 0.0001 M EDTA, pH 8.0) and human hemoglobin.[47] The latter study followed the changes in absorbance at 450 nm to monitor the release of the 2-nitro-5-mercaptobenzoic acid. The molar extinction coefficients obtained at 450 nm were 5550 M^{-1} cm^{-1} (pH 6.0); 6510 M^{-1} cm^{-1} (pH 7.0); 6810 M^{-1} cm^{-1} (pH 8.0); 6940 M^{-1} cm^{-1} (pH 9.0) and 7010 M^{-1} cm^{-1} (pH 9.5). The synthesis of a selenium analogue of this class of reagents, 6,6-diselenobis-(3-nitrobenzoic acid), has been reported.[48] The selenium-containing reagent has the same reaction characteristics as the sulfur-containing compound in terms of specificity of reaction with cysteinyl residues in proteins. The reaction is monitored by spectroscopy following the release of 6-seleno-3-nitrobenzoate which has a maximum at 432 nm (Figure 41). The extinction coefficient for the 6-seleno-3-nitrobenzoate anion varies slightly from 9,532 M^{-1}cm^{-1} (with excess reagent) to 10,200 M^{-1}cm^{-1} (with either excess cysteine or excess β-mercaptoethanol).

The above reagents (5,5'-dithiobis-(1-nitrobenzoic acid), 4,4'-dithiodipyridine, etc.) utilize mixed disulfide formation with reagent to obtain modification at cysteinyl residues. There are several other examples of this approach which are worth further consideration. Cystine or cystamine have proved effective in the modification of guanylate cyclase[49] as shown in Figure 42. Note the ready reversibility of the modification on the addition of dithiothreitol. Kaiser and co-workers[50] introduced methyl 3-nitro-2-pyridyl disulfide and methyl 2-pyridyl disulfide. Both of these reagents modify sulfhydryl groups forming the thiomethyl derivative. Figure 43 shows the spectra of methyl 3-nitro-2-pyridyl disulfide (NPySSMe) and 3-nitro-2-pyridone (NPySH) with determinations of the latter compared at several different conditions of pH. The spectrum of 3-nitro-2-pyridone is pH dependent. There is an isosbestic point at 310.4 nm which can be used to determine the extent of the reaction of methyl 3-nitro-2-pyridyl disulfide with sulfhydryl groups. Similar spectral studies for methyl 2-pyridyl disulfide (PySSMe) and 2-thio-pyridone (PySH) are shown in Figure 44. The difference spectrum obtained does not show the pH dependence of the nitropyridyl derivative (see Figure 43). At 343 nm, the change in extinction coefficient is 7,060 M^{-1} cm^{-1}. Confirmation of *S*-methyl cysteine and 2-thiopyridone as the reaction products from

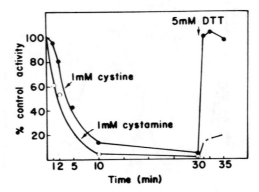

FIGURE 42.   The effect of mixed disulfide formation
on guanylate cyclase activity. Guanylate cylase was incubated with either 1 mM cystine or 1 mM cystamine
in 0.02 M Tris-HCl, pH 7.6, 1 mM dithiothreitol, 10%
sucrose. After 30 min of incubation, dithiothreitol was
added to a final concentration of 5 mM. (From Brandwein, J., Lewicki, J., and Murad, F., *J. Biol. Chem.*,
256, 2958, 1981. With permission.)

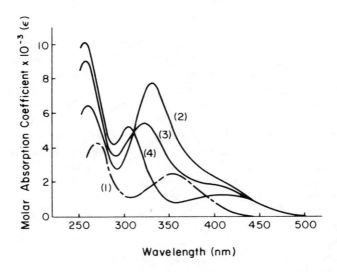

FIGURE 43.   The UV absorption spectra of methyl-3-nitro-2-pyridyl
disulfide (NPySSMe) and 3-nitro-2-pyridone (NPySH) in 0.050 M sodium phosphate: NPySSMe at pH 4.8 (curve 1), NPySH at pH 4.0,
4.6 (curve 2), pH 6.4 (curve 3), and pH 8.4, 8.8 (curve 4). (From
Kimura, T., Matsueda, R., Nakagawa, Y., and Kaiser, E. T., *Analyt.
Biochem.*, 122, 274, 1982. With permission.)

L-cysteine and methyl-2-pyridyl disulfide was obtained from NMR spectroscopy (Figure
45). The time course for the reaction of methyl-2-pyridyl disulfide with glutathione or papain
is shown in Figure 46. Note the differences in the rate of the reaction of methyl-2-pyridyl
disulfide with the cystinyl residue in glutathione and papain.

The reaction of the single cysteinyl residue in albumin has been studied by Pederson and
Jacobsen.[51] The suggested reaction mechanism and the spectra of 2,2′-dithiodipyridine and
2-thiopyridinone are shown in Figure 47. The extinction coefficient (7,600 M^{-1} cm^{-1}) of

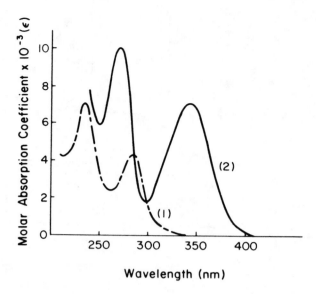

FIGURE 44. The UV absorption spectra of methyl-2-pyridyl disulfide (PySSMe) and 2-thiopyridone (PySH) in 0.050 M sodium phosphate, pH 7.5. Curve 1 is that for PySSMe and curve 2 for PySH. (From Kimura, T., Matsueda, R., Nakagawa, Y., and Kaiser, E. T., *Analyt. Biochem.*, 122, 274, 1982. With permission.)

the 2-thiopyridinone at 343 nm is relatively stable from pH 3 to pH 8.0. Above pH 8.0 there is a marked decrease reflecting the loss of a proton. Reaction with the sulfhydryl group in the protein clearly proceeds more rapidly at alkaline pH.

p-Hydroxy mercuribenzoate continues to be of use for the modification of sulfhydryl groups in proteins. The reagent is obtained as p-chloromercuribenzoate but is instantaneously converted to the hydroxy derivative in aqueous solution. This reagent was originally described by Boyer.[52] The absorbance change at 255 nm upon modification is 6200 M^{-1} cm^{-1} at pH 4.6 and 7600 M^{-1} cm^{-1} at pH 7.0. Several examples of the use of mercuribenzoates are discussed in further detail. Katz[53] has examined changes in the reactivity of cysteinyl residues in actin upon the addition of certain divalent cations. In these experiments, the extent of reaction was determined by changes in the absorbance at 255 nm. In Figure 48, it can be seen that calcium ions apparently decrease the reactivity of cysteinyl residues in actin while other divalent cations increase reactivity (Figure 49). Bai and Hayashi[54] have examined the reaction of organic mercurials with yeast carboxypeptidase (carboxypeptidase Y). The titration of the catalytically essential cysteinyl residue in carboxypeptidase Y with p-hydroxymercuribenzoate is shown in Figure 50. The pH dependence for this reaction is shown in Figure 51 and is similar to pH dependence observed with the hydrolysis of peptide substrates. Treatment of the modified enzyme with millimolar cysteine resulted in virtually complete recovery of catalytic activity.

2-Chloromercuri-4-nitrophenol is a compound related to the organic mercurial described above. It has proved useful as a "reporter" group in the study of microenvironmental changes in the modified protein. An excellent example of this is provided from the studies of Marshall and Cohen[55] on the properties of ornithine transcarbamylase modified with 2-chloromercuri-4-nitrophenol. The enzyme from *S. faecalis* was modified in 0.1 M MOPS, 0.1 M KCl, pH 7.5 using changes in absorbance at 403 nm to follow the extent of modification. The pH dependence of the spectrum of the modified *S. faecalis* enzyme is shown in Figure 52. The bovine enzyme is carboxamido methylated on a nonessential sulfhydryl group before reaction

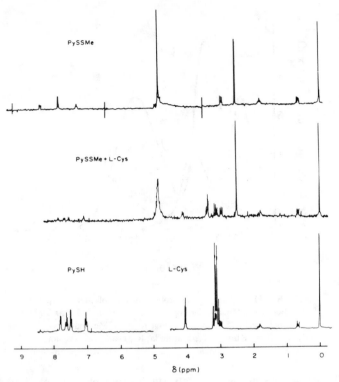

FIGURE 45.   The nuclear magnetic resonance spectra of methyl-2-pyr-
idyl disulfide (PySSMe), a mixture of PySSMe and cysteine (L-Cys) and
a mixture of 2-thiopyridone (PySH) and cysteine (L-Cys). (From Kimura,
T., Matsueda, R., Nakagawa, Y., and Kaiser, E. T., (*Analyt. Biochem.*,
122, 274, 1982. With permission.)

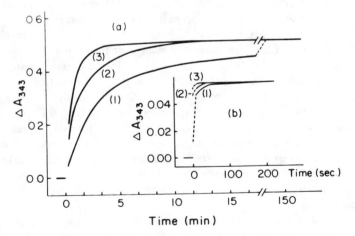

FIGURE 46.   Time course studies of the formation of 2-thiopyridone resulting from the reaction of methyl-2-pyridyl disulfide (PySSMe) with glutathione (a) or papain (b). In the experiments with glutathione (a), the
concentration of glutathione was 71.7 μ*M* and the ratio of PySSMe to glutathione was 1.04 (1), 1.57 (2), and 3.13 (3). In the experiments with papain
(b), the concentration of papain was 8.29 μ*M* and the ration of PySSMe to
papain was 1.02 (1), 1.70 (2), and 3.41(3). (From Kimura, T., Matsueda,
R., Nakagawa, Y., and Kaiser, E. T., *Analyt. Biochem.*, 122, 274, 1982.
With permission.)

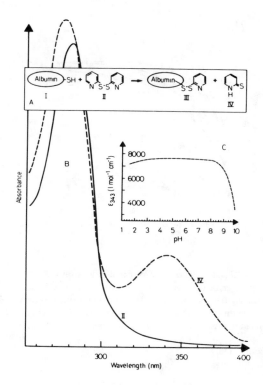

FIGURE 47. (A) A scheme for the reaction of 2,2'-dithiopyridine and mercaptalbumin; (B) The UV absorption spectra for 2,2'-dithiodipyridine(solid line, II) and 2-thiopyridinone (dashed line, IV) in 0.1 *M* sodium phosphate, pH 7.0; (C) The molar extinction coefficient at 343 nm for 2-thiopyridinone as a function of pH. (From Pedersen, A. O. and Jacobsen, J., *Eur. J. Biochem.*, 106, 291, 1980. With permission.)

with the organic mercurial. Modification of the bovine enzyme with 2-chloromercuri-4-nitrophenol is performed in 0.020 *M* MOPS, 0.1 *M* KCl, pH 7.11 at 25°C. The modification was followed by the change in absorbance at 405 nm. The effect of pH on the spectrum of the modified bovine enzyme is shown in Figure 53. Baines and Brocklehurst[56] have reported the synthesis and characterization of 2-(2'-pyridylmercapto) mercuri-4-nitrophenol, a reagent which does have certain advantages. In particular, the spectral changes occurring on modification (Figure 54) permit the more facile *in situ* determination of the extent of reaction.

A number of other modifications of sulfhydryl groups have proved useful. *O*-methylisourea reacts with cysteinyl residues to form the *S*-methyl derivative (Figure 55). Shafer's laboratory has reported on the modification of papain with *O*-methylisourea[57] as shown in Figure 56. Carbodiimides have been demonstrated to react with sulfhydryl groups.[58,59] Cyanate also can modify sulfhydryl groups as shown in Table 3.[60] The carbamoyl derivative of cysteine is stable at acid pH but rapidly decomposes at alkaline pH.

Chloro-7-nitrobenzo-2-oxa-1,3-diazole (Figure 57) is a reagent developed for the modification of amino groups.[61] It has also found application in the modification of sulfhydryl groups and is useful in that it introduces a fluorescent probe.[62-66] Nitta and co-workers[65] have noted that there are other possible reaction products of 4-chloro-7-nitrobenzo-2-oxa-1,3-diazole (4-chloro-7-nitrobenzofurazan; Nbf-Cl) including the possibility of reaction products with sulfhydryl groups. The modification of the sulfhydryl group with concomitant reaction at the 4-position yields a derivative with molar absorption coefficient of 13,000

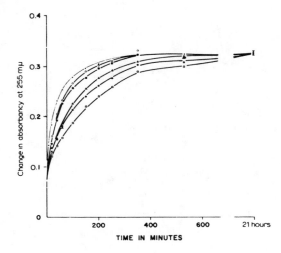

FIGURE 48.   The effect of calcium ions on the reaction of
p-chloromercuribenzoate with sulfhydryl groups in actin. Increasing the concentration of calcium ions decreases the rate of
the modification of the sulfhydryl groups. The extent of modification was determined by spectral analysis at 255 nm. (From
Katz, A. M., *Biochim. Biophys. Acta,* 71, 397, 1963. With
permission.)

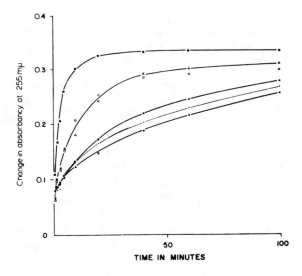

FIGURE 49.   The effect of several divalent cations on the
reactivity of sulfhydryl groups in actin as determined by reaction
with *p*-chloromercuribenzoate. Zinc ions (■), cobalt ions (△),
nickel ions (□) all stimulate reactivity while magnesium (○)
has lesser effect and manganese (●) inhibits when compared
to the control experiment (·). (From Katz, A. M., *Biochim.
Biophys. Acta,* 71, 397, 1963. (With permission.)

(Figure 58).[66] The reaction of 4-chloro-7-nitrobenzo-2-oxa-1,3-diazole with sulfhydryl groups
in glutathione reductase and lipoamide dehydrogenase has also been reported.[67] Nitta and
co-workers[65] have examined the chemistry of the reaction of 4-chloro-7-nitrobenzo-2-oxa-

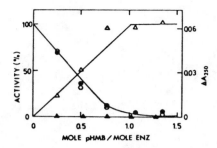

FIGURE 50. The spectrophotometric titration of native and DIP-carboxypeptidase Y with *p*-hydroxymercuribenzoate (*p*-HMB). The reaction with the protein was accomplished in 0.08 *M* sodium phosphate, pH 7.0. Increments of *p*-HMB were added to the protein preparations and allowed to stand for 20 min at which time absorbance at 250 nm was determined as catalytic activity (Z-Phe-Leu or Ac-Phe-OEt). The open circles indicate activity toward Ac-Phe-OEt, the closed circles indicate activity toward Z-Phe-Leu, the open triangles represent the absorbance at 250 nm of the native carboxypeptidase Y, and the closed triangles the absorbance at 250 nm of carboxypeptidase Y previously reacted with diisopropylphosphorofluoridate (DIP-carboxypeptidase Y). (From Bai, Y. and Hayashi, R., *J. Biol. Chem.*, 254, 8473, 1979. With permission.)

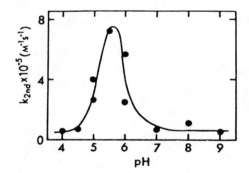

FIGURE 51. The pH dependence for the inactivation of carboxypeptidase Y by phenylmercuric acetate. Buffers used were 0.05 *M* sodium acetate (pH 4—6), 0.05 *M* sodium phosphate (pH 6—8), and 0.05 *M* sodium borate (pH 9.0). (From Bai, Y. and Hayashi, R., *J. Biol. Chem.*, 254, 8473, 1979. With permission.)

1,3-diazole with model sulfhydryl compounds in some detail. Figure 59 shows the dependence of the first-order rate constant on 4-chloro-7-nitrobenzo-2-oxa-1,3-diazole (Nbf-Cl) concentration. It is noted that the reaction rates are a measure of product formation rather than reagent disappearance. This is seen more clearly in Figure 60. The spectra of the reagent (Nbf-Cl) and product (4-(2'-hydroxy-ethylthio)-7-nitrobenzofuran; Nbf-OHEtS) are presented in Figure 61. These data should be compared with that presented in Figure 62. Note the dependence of the difference spectra on sulfhydryl concentration and pH/solvent species. The data in Figure 62, part I were obtained in triethanolamine, pH 7.5 with an approximate 20-fold molar excess of β-mercaptoethanol while that in part II was obtained with an approximate 200-fold molar excess. Experimental series part III was performed with a 20-fold molar excess of β-mercaptoethanol in sodium citrate buffer, pH 5.0. Similar spectral changes are seen with dithiothreitol in Figure 63. The data in experimental series I were obtained at an Nbf-Cl concentration of 0.148 n*M* and a dithiothreitol concentration of 0.0097

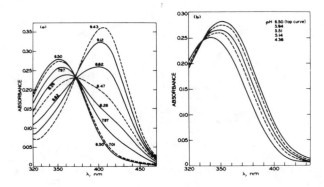

FIGURE 52.   The effect of pH on the spectrum of the 2-chloro-
mercuri-4-nitrophenol derivate of ornithine transcarbamylase. Panel
a represents the results obtained in the pH range of 9.43 to 6.50.
Panel b represents the results obtained in pH the range of 6.5 (top
curve) to 4.36 (bottom curve). (From Marshall, M. and Cohen, P.
P., *J. Biol. Chem.*, 225, 7296, 1980. With permission.)

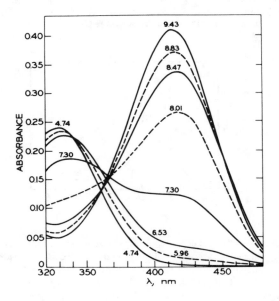

FIGURE 53.   The effect of pH on the UV absorption
spectrum of the 2-chloromercuri-4-nitrophenol derivative
of the monocarboxamidomethyl bovine liver ornithine
transcarbamylase. (From Marshall, M. and Cohen, P. P.,
J. Biol. Chem., 255, 7296, 1980. With permission.)

mM while the data in experimental series III were obtained at an Nbf-Cl concentration of
0.0742 mM and a dithiothreitol concentration of 9.38 mM. The variety of potential products
in this reaction require that considerable caution be given to the interpretation of spectra
data obtained with this reagent.

The reaction of cysteinyl residues with acrylonitrile has also been reported.[68,69] The reaction
at neutral to basic pH initially yields the *S*-cyanoethyl derivative, which is converted to the
S-carboxyethyl derivative (Figure 64).

The modification of cysteinyl residues in proteins with 2-bromoethane sulfonate has been

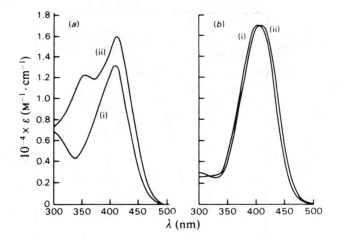

FIGURE 54. The UV absorption spectra of (a) 2-(2'-pyridylmer-capto)mercuri-4-nitrophenol and (b) 2-chloromercuri-4-nitrophenol before (i) and after (ii) the addition of an excess of β-mercaptoethanol. The solvent was sodium/potassium phosphate (ionic strength = 0.1) containing 13% (v/v) ethanol. The reaction of the mercu-rial(concentration = 10 μM) with β-mercaptoethanol (concentration = 45 mM) was complete after 5 min of reaction. (From Baines, B. S. and Brocklehurst, K., *Biochem. J.*, 179, 701, 1979. With permission.)

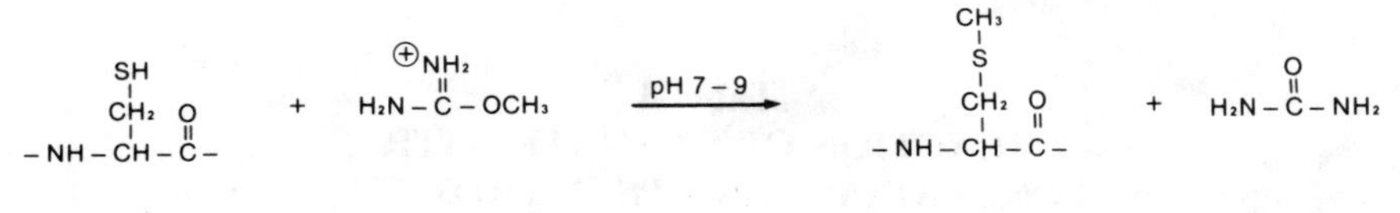

FIGURE 55. The reaction of *O*-methylisourea with cysteine.

reported.[70] This derivatization procedure was developed in response to a need for a strongly hydrophilic substituent in samples for the Edman degradation. The modification time is longer than for the corresponding carboxymethyl derivatives, taking 12 hr for lysozyme, 24 hr for insulin, and 48 hr for glutathione. This derivative has considerable utility since the *S*-sulfoethylated lysozyme derivative is soluble between pH 5.0 and 10.0 while the *S*-carboxymethylated derivative is not. This procedure has potential for primary structure analysis.

A derivatization procedure that has proved useful in the primary structure analysis of protein has been the reaction of ethyleneimine (Figure 65) with sulfhydryl groups in proteins.[71] This reaction produces *S*-aminoethyl cysteine which provides an additional point of tryptic cleavage in proteins.[72] With bovine pancreatic ribonuclease A, with a 1/100 ratio of trypsin at pH 8.0, 83% cleavage of arginyl and lysyl bonds was obtained while 56% cleavage was obtained at *S*-aminoethyl cysteine.

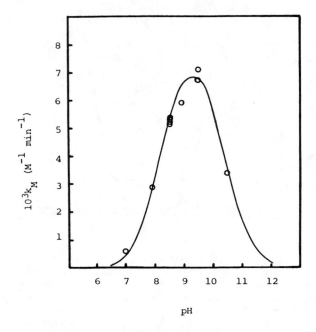

FIGURE 56. The pH dependence for the second-order rate constant for the reaction of papain with 1.0 *M O*-methylisourea. (From Banks, T. E. and Shafer, J. A., *Biochemistry*, 11, 110, 1972. With permission.)

Table 3
REACTION OF CYANATE WITH FUNCTIONAL GROUPS IN PROTEINS

Functional group	pKa	M^{-1} min^{-1}
Alpha amino	7.8—8.2	1.4×10^{-1}
Epsilon amino	10.5-10.8	2×10^{-3}
Sulfhydryl	8.3—8.5	4.0
Imidazole	7.0—7.2	1.8×10^{-1}

From Stark, G. R., *Meth. Enzymol.*, 11, 590, 1967. With permission.

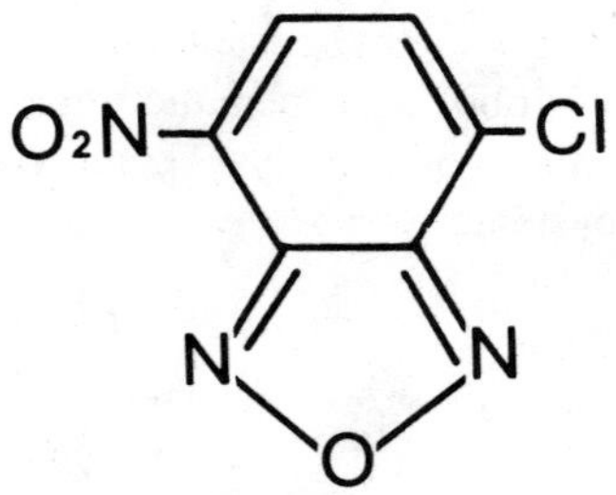

FIGURE 57. The structure of 4-chloro-7-nitrobenzo-2-oxa-1,3-diazole (4-chloro-7-nitrobenzofurazan, NBF-Cl).

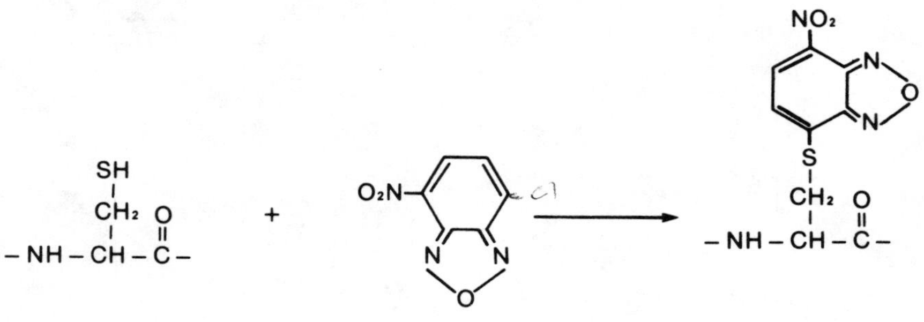

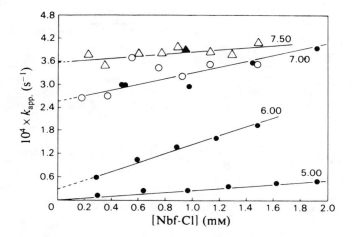

FIGURE 58.   The reaction of 4-chloro-7-nitrobenzo-2-oxa-1,3-diazole with cysteine.

FIGURE 59.   The dependence of the apparent first-order rate constant on 4-chloro-7-nitrobenzo-2-oxa-1,3-diazole concentration on reaction with β-mercaptoethanol at various pH values (pH 7.5 (▲), 7.0, 6.0, 5.0 (●); 0.05 M sodium citrate; pH 7.5 (△), 0.05 M triethanolamine HCl/NaOH, and 0.05 M potassium/sodium phosphate, pH 7.0 (○). (From Nitta, K., Bratcher, S. C., and Kronman, M. J., *Biochem. J.*, 177, 385, 1979. With permission.)

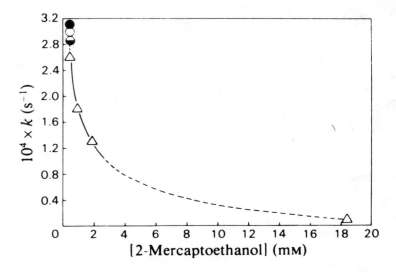

FIGURE 60.   The effect of β-mercaptoethanol concentration on the apparent first-order rate constant for reaction with 4-chloro-7-nitrobenzo-2-oxa-1,3-diazole. The solvent was 0.05 *M* triethanolamine HCl/NaOH, pH 7.5. (From Nitta, K., Bratcher, S. C., and Kronman, M. J., *Biochem. J.*, 177, 385, 1979. With permission.)

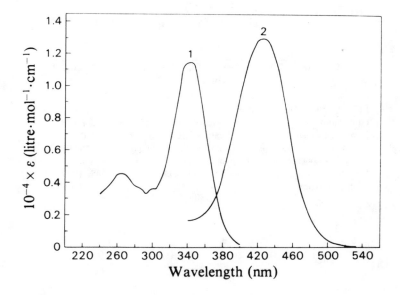

FIGURE 61.   The UV absorption spectra of 4-chloro-7-nitrobenzo-2-oxa-1,3-diazole(Nbf-Cl) (curve 1) and the reaction product between Nbf-Cl and β-mercaptoethanol (curve 2). The solvent was water adjusted to pH 7.1. (From Nitta, K., Bratcher, S. C., and Kronman, M. J., *Biochem. J.*, 177, 385, 1979. With permission.)

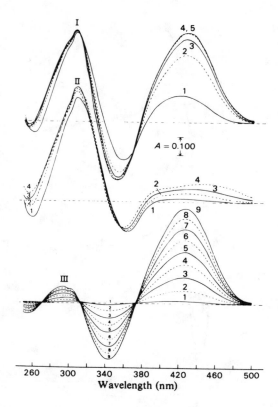

FIGURE 62.   Different spectra of mixtures of 4-chloro-7-nitrobenzofurazan (Nbf-Cl) and β-mercaptoethanol at ambient temperature. The spectra were scanned at a rate of 30 nm/min from high to low wavelength. The times given below are for the initial wavelength in the scan. The solution containing the Nbf-Cl + β-mercaptoethanol was in the sample beam. The spectra in set I were obtained in 0.05 M triethanolamine HCl/NaOH, pH 7.5 at an Nbf-Cl concentration of 0.102 mM and a β-mercaptoethanol concentration of 2.28 mM. The times for the individual curves in set I were 2 min (curve 1), 12 min (curve 2), 22 min (curve 3), 42 min (curve 4), and 72 min (curve 5). The spectra in set II were obtained in 0.05 M triethanolamine HCl/NaOH buffer, pH 7.5, at an Nbf-Cl concentration of 0.104 mM and a β-mercaptoethanol concentration of 22.8 mM. The times for the individual curves in set II were 3 min(curve 1), 15 min(curve 2), 35 min(curve 3), and 60 min(curve 4). The spectra in set III were obtained in 0.05 M sodium citrate/citric acid, pH 5.0 at an Nbf-Cl concentration of 0.112 mM. The times for the individual curves of set III were 3 min(curve 1), 27 min(curve 2), 57 min(curve 3), 87 min(curve 4), 117 min(curve 5), 147 min(curve 6), 177 min(curve 7), 207 min(curve 8), and 237 min(curve 9). (From Nitta, K., Bratcher, S. C., and Kronman, M. J., *Biochem. J.,* 177, 385, 1979. With permission.)

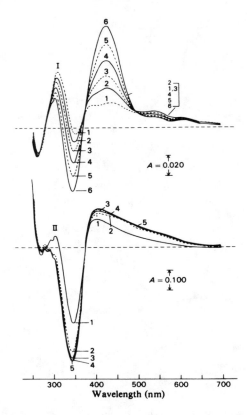

FIGURE 63.   Different spectra of mixtures of 4-chloro-7-nitrobenzofurazan(Nbf-Cl) and dithiothreitol at ambient temperature. Spectra were scanned at rate of 30 nm/min from high to low wavelength. The times given below were for the initial wavelength in the scan. The solution of Nbf-Cl + thiol compound was in the sample beam. The buffer used was 0.05 N triethanolamine HCl/NaOH (pH 7.5). The spectra of set I were obtained at an Nbf-Cl concentration of 0.148 mM and a dithiothreitol concentration of 9.97 μM at 3 min (curve 1), 13 min (curve 2), 23 min (curve 3), 38 min (curve 4), 58 min (curve 5), and 88 min (curve 6). The spectra in set II were obtained at an Nbf-Cl concentration of 9.389 mM at 2 min (curve 1), 12 min (curve 2), 22 min (curve 3), 42 min (curve 4), and 72 min (curve 5). (From Nitta, K., Bratcher, S. C., and Kronman, M. J., *Biochem. J.*, 177, 385, 1979. With permission.)

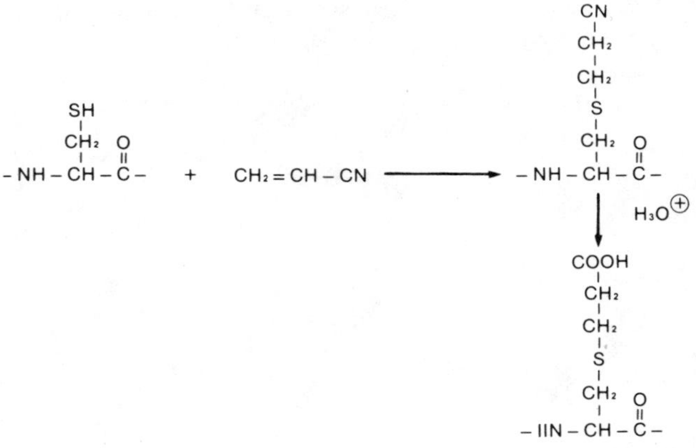

FIGURE 64.   The reaction of acrylonitrile with cysteine.

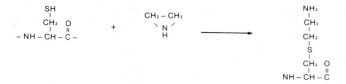

FIGURE 65.   The reaction of ethyleneimine with cysteine.

REFERENCES

1. **Liu, T.-Y.,** The role of sulfur in proteins, *The Proteins,* Vol. 3, 3rd ed., Neurath, H. and Hill, R. L., Eds., Academic Press, New York, 1977, 240.
2. **Gerwin, B. I.,** Properties of the single sulfhydryl group of streptococcal proteinase. A comparison of the rates of alkylation by chloroacetic acid and chloroacetamide, *J. Biol. Chem.,* 242, 451, 1967.
3. **Liu, T. Y.,** Demonstration of the presence of a histidine residue at the active site of streptococcal proteinase, *J. Biol. Chem.,* 242, 4029, 1967.
4. **Pihl, A. and Lange, R.,** The interaction of oxidized glutathione, cystamine monosulfoxide, and tetrathionate with the -SH groups of rabbit muscle D-glyceraldehyde 3-phosphate dehydrogenase, *J. Biol. Chem.,* 237, 1356, 1962.
5. **Chaiken, I. M. and Smith, E. L.,** Reaction of chloroacetamide with the sulfhydryl group of papain, *J. Biol. Chem.,* 244, 5087, 1969.
6. **Chaiken, I. M. and Smith, E. L.,** Reaction of the sulfhydryl group of papain with chloroacetic acid, *J. Biol. Chem.,* 244, 5095, 1969.
7. **Jörnvall, H., Fowler, A. V., and Zabin, I.,** Probe of β-galactosidase structure with iodoacetate. Differential reactivity of thiol groups in wild-type and mutant forms of β-galactosidase, *Biochemistry,* 17, 5160, 1978.
8. **Kalimi, M. and Love, K.,** Role of chemical reagents in the activation of rat hepatic glucocorticoid-receptor complex, *J. Biol. Chem.,* 255, 4687, 1980.
9. **Kallis, G.-B. and Holmgren, A.,** Differential reactivity of the functional sulfhydryl groups of cysteine-32 and cysteine-35 present in the reduced form of thioredoxin from *Escherichia coli, J. Biol. Chem.,* 255, 10261, 1980.
10. **Mikami, B., Aibara, S., and Morita, Y.,** Chemical modification of sulfhydryl groups in soybean β-amylase, *J. Biochem.,* 88, 103, 1980.
11. **Hempel, J. D. and Pietruszko, R.,** Selective chemical modification of human liver aldehyde and dehydrogenases E_1 and E_2 by iodoacetamide, *J. Biol. Chem.,* 256, 10889, 1981.
12. **Crestfield, A. M., Moore, S., and Stein, W. H.,** The preparation and enzymatic hydrolysis of reduced and S-carboxymethylated proteins, *J. Biol. Chem.,* 238, 622, 1963.
13. **Friedman, M., Krull, L. H., and Cavins, J. F.,** The chromatographic determination of cystine and cysteine residues in proteins as S-β-(4-pyridyl-ethyl) cysteine, *J. Biol. Chem.,* 245, 3868, 1970.
14. **Mak, A. S. and Jones, B. L.,** Application of S-pyridylethylation of cysteine to the sequence analysis of proteins, *Analyt. Biochem.,* 84, 432, 1978.
15. **Huestis, W. H. and Raftery, M. A.,** A study of cooperative interactions in hemoglobin using fluorine nuclear magnetic resonance, *Biochemistry,* 11, 1648, 1972.
16. **Huestis, W. H. and Raftery, M. A.,** Conformation and cooperativity in hemoglobin, *Biochemistry,* 14, 1886, 1975.
17. **Huestis, W. H. and Raftery, M. A.,** Bromotrifluoroacetone alkylates hemoglobin at cysteine β93, *Biochem. Biophys. Res. Commun.,* 81, 892, 1978.
18. **Dahl, K. S. and McKinley-McKee, J. S.,** The reactivity of affinity labels: a kinetic study of the reaction of alkyl halides with thiolate anions — a model reaction for protein alkylation, *Bioorganic Chem.,* 10, 329, 1981.
19. **Holowka, D.,** N-chloroacetyl-[^{125}I] iodotyramine: an alkylating agent with high specific activity, *Analyt. Biochem.,* 117, 390, 1981.
20. **Yankeelov, J. A., Jr. and Jolley, C. J.,** α-Bromo-β(5-imidazoyl) propionic acid and its reaction with cysteine, *Biochemistry,* 11, 159, 1972.
21. **Jolley, C. J. and Yankeelov, J. A., Jr.,** Reaction of papain with α-bromo-β-(5-imidazoyl) propionic acid, *Biochemistry,* 11, 164, 1972.

22. **Paterson, M. C., Norton, I. L., and Hartman, F. C.,** Haloacetol phosphates. Selective alkylation of sulfhydryl groups of rabbit muscle aldolase by chloroacetol phosphate, *Biochemistry,* 11, 2070, 1972.

23. **Salter, D. W., Norton, I. L., and Hartman, F. C.,** Haloacetol phosphates. Identification of the sulfhydryl group of rabbit muscle aldolase alkylated by chloroacetol phosphate, *Biochemistry,* 12, 1, 1973.

24. **Hartman, F. C.,** Haloacetol phosphates. Potential active-site reagents for aldolase, triosephosphate isomerase and glycerophosphate dehydrogenase. I. Preparation and properties, *Biochemistry,* 9, 1776, 1970.

25. **Gregory, J. D.,** The stability of *N*-ethylmaleimide and its reaction with sulfhydryl groups, *J. Am. Chem. Soc.,* 77, 3922, 1955.

26. **Leslie, J.,** Spectral shifts in the reaction of *N*-ethylmaleimide with proteins, *Analyt. Biochem.,* 10, 162, 1965.

27. **Gorin, G., Martic, P. A., and Doughty, G.,** Kinetics of the reaction of *N*-ethylmaleimide with cysteine and some congeners, *Arch. Biochem. Biophys.,* 115, 593, 1966.

28. **Smyth, D. G., Blumenfeld, O. O., and Konigsberg, W.,** Reaction of *N*-ethylmaleimide with peptides and amino acids, *Biochem. J.,* 91, 589, 1964.

29. **Portanova, J. P. and Shrift, A.,** Usefulness of *N*-ethylmaleimide in the identification of ^{75}Se-labeled selenocysteine, *J. Chromatog.,* 139, 391, 1977.

30. **Gehring, H. and Christen, P.,** A diagonal procedure for isolating sulfhydryl peptides alkylated with *N*-ethylmaleimide, *Analyt. Biochem.,* 107, 358, 1980.

31. **Brown, R. D. and Matthews, K. S.,** Chemical modification of lactose repressor proteins using *N*-substituted maleimides, *J. Biol. Chem.,* 254, 5128, 1979.

32. **Brown, R. D. and Matthews, K. S.,** Spectral studies on *Lac* repressor modified with *N*-substituted maleimide probes, *J. Biol. Chem.,* 254, 5135, 1979.

33. **Le-Quoc, K., Le-Quoc, D., and Gaudemer, Y.,** Evidence for the existence of two classes of sulfhydryl groups essential for membrane-bound succinate dehydrogenase activity, *Biochemistry,* 20, 1705, 1981.

34. **Yamamoto, K., Sekine, T., and Kanaoka, Y.,** Fluorescent thiol reagents. XII. Fluorescent tracer method for protein SH groups using *N*-(7-dimethylamino-4-methyl coumarinyl) maleimide. An application to the proteins separated by SDS polyacrylamide gel electrophoresis, *Analyt. Biochem.,* 79, 83, 1977.

35. **Yamamoto, K., Okamoto, Y., and Sekine, T.,** Fluorescent tracer method for protein SH groups. II, *Analyt. Biochem.,* 84, 313, 1978.

36. **Yamamoto, K. and Sekine, T.,** Fluorescent tracer method for protein SH groups. III. Use of *N*-(7-dimethylamino-4-methylcoumarinyl) maleimide as a tracer of cysteine-containing peptides, *Analyt. Biochem.,* 90, 300, 1978.

37. **Degani, Y. and Degani, C.,** Subunit-selective chemical modifications of creatine kinase. Evidence for asymmetrical association of the subunits, *Biochemistry,* 18, 5917, 1979.

38. **Ellman, G. L.,** Tissue sulfhydryl groups, *Arch. Biochem. Biophys.,* 82, 70, 1959.

39. **Pecci, L., Cannella, C., Pensa, B., Costa, M., and Cavallini, D.,** Cyanylation of rhodanese by 2-nitro-5-thiocyanobenzoic acid, *Biochem. Biophys. Acta,* 623, 348, 1980.

40. **Marshall, M. and Cohen, P. P.,** Ornithine transcarbamylases. Ordering of *S*-cyanopeptides and location of characteristically reactive cysteinyl residues within the sequence, *J. Biol. Chem.,* 255, 7287, 1980.

41. **Ogilvie, J. W.,** Cleavage of phosphofructokinase at *S*-cyanylated cysteine residues, *Biochim. Biophys. Acta,* 622, 277, 1980.

42. **Kindman, L. A. and Jencks, W. P.,** Modification and inactivation of CoA transferase by 2-nitro-5-(thiocyanato) benzoate, *Biochemistry,* 20, 5183, 1981.

43. **Talgoy, M. M., Bell, A. W., and Duckworth, H. W.,** The reactions of *Escherichia coli* citrate synthase with the sulfhydryl reagents 5,5'-dithiobis-(2-nitrobenzoic acid) and 4,4'-dithiodipyridine, *Can. J. Biochem.,* 57, 822, 1979.

44. **Soper, T. S., Jones, W. M., and Manning, J. M.,** Effects of substrates on the selective modification of the cysteinyl residues of D-amino acid transaminase, *J. Biol. Chem.,* 254, 10901, 1979.

45. **Lukas, R. J. and Bennett, E. L.,** Chemical modification and reactivity of sulfhydryls and disulfides of rat brain nicotinic-like acetylcholine receptors, *J. Biol. Chem.,* 255, 5573, 1980.

46. **Cockle, S. A., Epand, R. M., Stollery, J. G., and Moscarello, M. A.,** Nature of the cysteinyl residues in lipophilin from human myelin, *J. Biol. Chem.,* 255, 9182, 1980.

47. **Hallaway, B. E., Hedlund, B. E., and Benson, E. S.,** Studies of the effect of reagent and protein charges on reactivity of the β93 sulfhydryl group of human hemoglobin using selected mutations, *Arch. Biochem. Biophys.,* 203, 332, 1980.

48. **Luthra, M. P., Dunlap, R. B., and Odom, J. D.,** Characterization of a new sulfhydryl group reagent: 6,6'-diselenobis-(3-nitrobenzoic acid), a selenium analog of Ellman's reagent, *Analyt. Biochem.,* 117, 94, 1981.

49. **Brandwein, H. J., Lewicki, J. A., and Murad, F.,** Reversible inactivation of guanylate cyclase by mixed disulfide formation, *J. Biol. Chem.,* 256, 2958, 1981.

50. **Kimura, T., Matsueda, R., Nakagawa, Y., and Kaiser, E. T.,** New reagents for the introduction of the thiomethyl group at sulfhydryl residues of proteins with concomitant spectrophotometric titration of the sulfhydryl: methyl 3-nitro-2-pyridyl disulfide and methyl 2-pyridyl disulfide, *Analyt. Biochem.,* 122, 274, 1982.
51. **Pedersen, A. O. and Jacobsen, J.,** Reactivity of the thiol group in human and bovine albumin at pH 3-9, as measured by exchange with 2,2'-dithiodipyridine, *Eur. J. Biochem.,* 106, 291, 1980.
52. **Boyer, P. D.,** Spectrophotometric study of the reaction of protein sulfhydryl groups with organic mercurials, *J. Am. Chem. Soc.,* 76, 4331, 1954.
53. **Katz, A. M.,** The influence of cations on the reactivity of the sulfhydryl groups of actin, *Biochim. Biophys. Acta,* 71, 397, 1963.
54. **Bai, Y. and Hayashi, R.,** Properties of the single sulfhydryl group of carboxypeptidase Y. Effects of alkyl and aromatic mercurials on activities toward various synthetic substrates, *J. Biol. Chem.,* 254, 8473, 1979.
55. **Marshall, M. and Cohen, P. P.,** The essential sulfhydryl group of ornithine transcarbamylases-pH dependence of the spectra of its 2-mercuri-4-nitrophenol derivative, *J. Biol. Chem.,* 255, 7296, 1980.
56. **Baines, B. S. and Brocklehurst, K.,** A thiol-labelling reagent and reactivity probe containing electrophilic mercury and a chromophoric leaving group, *Biochem. J.,* 179, 701, 1979.
57. **Banks, T. E. and Shafer, J. A.,** Inactivation of papain by S-methylation of its cysteinyl residue with *O*-methylisourea, *Biochemistry,* 11, 110, 1972.
58. **Perfetti, R. B., Anderson, C. D., and Hall, P. L.,** The chemical modification of papain with 1-ethyl-3-(3-dimethyl-amino propyl) carbodiimide, *Biochemistry,* 15, 1735, 1976.
59. **Carraway, K. L. and Triplett, R. B.,** Reaction of carbodiimides with proteins sulfhydryl groups, *Biochim. Biophys. Acta,* 200, 564, 1970.
60. **Stark, G.,** Modification of proteins with cyanate, *Meth. Enzymol.,* 11, 590, 1967.
61. **Ghosh, P. B. and Whitehouse, M. W.,** 7-Chloro-4-nitrobenzo-2-oxa-1,3-diazole: a new fluorigenic reagent for amino acids and other amines, *Biochem. J.,* 108, 155, 1968.
62. **Birkett, D. J., Price, N. C., Radda, G. K., and Salmon, A. G.,** The reactivity of SH groups with a fluorogenic reagent, *FEBS Lett.,* 6, 346, 1970.
63. **Birkett, D. J., Dwek, R. A., Radda, G. K., Richards, R. E., and Salmon, A. G.,** Probes for the conformational transitions of phosphorylase b. Effect of ligands studied by proton relaxation enhancement, fluorescence and chemical reactivities, *Eur. J. Biochem.,* 20, 494, 1971.
64. **Lad, P. M., Wolfman, N. M., and Hammes, G. G.,** Properties of rabbit muscle phosphofructokinase modified with 7-chloro-4-nitrobenzo-2-oxa-1,3-diazole, *Biochemistry,* 16, 4802, 1977.
65. **Nitta, K., Bratcher, S. C., and Kronman, M. J.,** Anomalous reaction of 4-chloro-7-nitrobenzofurazan with thiol compounds, *Biochem. J.,* 177, 385, 1979.
66. **Dwek, R. A., Radda, G. A., Richards, R. E., and Salmon, A. G.,** Probes for the conformational transitions of phosphorylase a. Effect of ligands studied by proton-relaxation enhancement, and chemical reactivities, *Eur. J. Biochem.,* 29, 509, 1972.
67. **Carlberg, I. and Mannervik, B.,** Interaction of 2,4,6-trinitrobenzenesulfonate and 4-chloro-7-nitrobenzo-2-oxa-1,3-diazole with the active sites of glutathione reductase and lipoamide dehydrogenase, *Acta Chem. Scand.,* B34, 144, 1980.
68. **Kalan, E. B., Neistadt, A., Weil, L., and Gordan, W. G.,** The determination of S-carboxyethylcysteine and the cyanoethylation of milk proteins, *Analyt. Biochem.,* 12, 488, 1965.
69. **Seibles, T. S. and Weil, L.,** Reduction and S-alkylation with acrylonitrile, *Meth. Enzymol.,* 11, 204, 1967.
70. **Niketic, V., Thomsen, J., and Kristiansen, K.,** Modification of cysteine residues with 2-bromoethane-sulfonate. The application of S-sulfoethylated peptides in automatic Edman degradation, *Eur. J. Biochem.,* 46, 547, 1974.
71. **Raftery, M. A. and Cole, R. D.,** On the aminoethylation of proteins, *J. Biol. Chem.,* 241, 3457, 1966.
72. **Plapp, B. V., Raftery, M. A., and Cole, R. D.,** The tryptic digestion of S-aminoethylated ribonuclease, *J. Biol. Chem.,* 242, 265, 1967.

Chapter 7

THE MODIFICATION OF CYSTINE — CLEAVAGE OF DISULFIDE BONDS

It is generally accepted that disulfide bonds contribute substantially to the maintenance of the tertiary structure of single chain proteins such as ribonuclease[1,2] as well as to maintain the structure of multi-chain proteins such as fibrinogen. The denaturation and cleavage of disulfide bonds are necessary for the enzymatic or chemical cleavage of peptide bonds for the production of fragments suitable for subsequent analysis.

There are also instances in which cleavage of specific disulfide bond(s) can provide useful information regarding the relationships between structure and function in a protein. Cleavage of the disulfide bond connecting the A and B chains in thrombin has been accomplished by Scheraga and co-workers[3] in experiments which showed that the A chain did not have a critical role in the catalytic activity of this enzyme.

There are several approaches to the cleavage of disulfide bonds in proteins. The majority of studies involve the cleavage of the disulfide bond of cystine to the free thiol group of cysteine by reduction. Reduction has been generally accomplished with a mild reducing agent such as β-mercaptoethanol. Dithiothreitol has been a useful reagent in the reduction of disulfide bonds in proteins[4] as introduced by Cleland. Dithiothreitol and the isomeric form, dithioerythritol, are each capable of the quantitative reduction of disulfide bonds in proteins. Furthermore, the oxidized form of dithiothreitol has an absorbance maximum at 283 nm ($\Delta\epsilon = 273$) which can be used to determine the extent of disulfide bond cleavage.[5] The UV spectra of dithiothreitol and oxidized dithiothreitol are shown in Figure 1. Insolubilized dihydrolipoic acid has also been proposed for use in the quantitative reduction of disulfide bonds.[6]

In most proteins, the free sulfhydryl groups (cysteine) derived from the reduction of cystine will, at alkaline pH, fairly rapidly undergo reoxidation to form the original disulfide bonds. This process can be accelerated by the sulfhydryl-disulfide interchange enzyme[2,7,8] or sulfhydryl oxidase.[9] Thus, it is necessary to "block" the new sulfhydryl groups by alkylation, arylation or reaction with dithionite (see Chapter 6).

A novel reaction has been developed by Neumann and co-workers[10] which allows for the reduction of disulfide bonds under mild conditions. Phosphorothioate reacts with disulfide bonds to yield the *S*-phosphorothioate derivatives.[10] The reaction proceeds optimally at alkaline pH (pH optimum 9.7) and the reaction product, *S*-phosphorothioate cysteine, has an absorbance maximum at 250 nm ($\epsilon = 631 \ M^{-1}cm^{-1}$) as shown in Figure 2. Phosphorothioate does not absorb at this wavelength. This reagent has been used to study the reactivity of disulfide bonds in ribonuclease.[11] In the absence of a denaturing agent (reaction conditions: tenfold molar acess of reagent, pH 9.0, 16 hr at 25°C), two specific disulfide bonds ($Cys_{65} - Cys_{72}$; $Cys_{58} - Cys_{110}$) are converted to phosphorothioate derivatives. The resultant derivative of ribonuclease is fully active in hydrolysis of RNA and has increased activity in the hydrolysis of cyclic cytidylic acid. The synthesis of radiolabeled phosphorothioate from either $[P^{32}]$ or $[S^{35}]$ thiophosphoryl chloride was reported in this study.

Light and co-workers have examined the susceptibility of disulfide bonds in trypsinogen to reduction.[12] At pH 9.0 (0.1 M sodium borate), a single disulfide bond ($Cys_{179} - Cys_{203}$) is cleaved in trypsinogen by 0.1 M NaBH$_4$. The resulting sulfhydryl groups are "blocked" by alkylation. The characterization of the modified protein has been performed by the same group.[13] The disulfide bond which is modified under these conditions is critical in establishing the structure of the primary specificity site in trypsin.

From the above studies, there is little doubt that the various disulfide bonds in a protein show different reactivity toward reducing agents. These differences in reactivity can be explored with various reagents and can be utilized with the aid of partial reduction followed

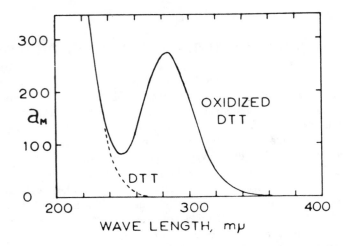

FIGURE 1.   The absorption spectra of dithiothreitol(DTT) and oxidized dithiothreitol (oxidized DTT) in aqueous solution. (From Cleland, W. W., *Biochemistry*, 3, 480, 1964. With permission.)

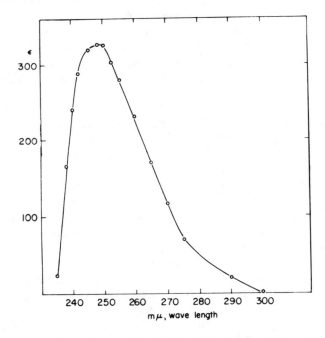

FIGURE 2.   The absorption spectrum of the reaction product formed from cystine and phosphorothioate(PS). (From Neumann, H. and Smith, R. L., *Arch. Biochem. Biophys.*, 122, 354, 1967. With permission.)

by alkylation with radiolabeled iodoacetate to determine the position of disulfide bonds in proteins.[14]

Gorin and Godwin[15] have reported that cystine can be quantitatively converted to cysteic acid by reaction with iodate in 0.1 to 1.0 *M* HCl. This reaction has been applied to insulin. The reaction product was not completely characterized, but given the relationship between iodate consumption and the cystine residues in insulin, the primary reaction is the oxidation

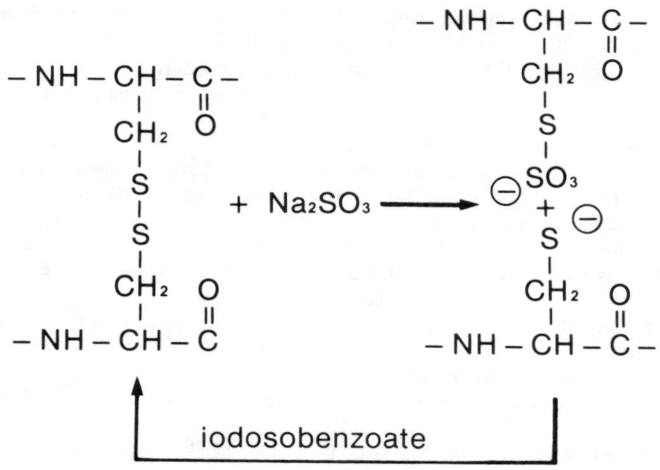

FIGURE 3.   The cleavage of disulfide bonds by sodium sulfite to form the *S*-sulfo derivative.

of disulfide bonds. This reaction was complete in 15 to 30 min. After longer periods of reaction, the iodination of tyrosine residues occurred.

Disulfide bonds are somewhat unstable at alkaline pH (pH $\geq$ 13.0). This has been examined by Donovan in some detail.[16] With protein-bound cystine, there is change in the spectrum with an increase in absorbance at 300 nm. This problem has been more recently studied by Florence.[17] This investigation presented evidence to suggest that cleavage of disulfide bonds in proteins by base proceeds via β-elimination to form dehydroalanine and a persulfide intermediate which can decompose to form several products.

The electrolytic reduction of proteins has been explored by Leach and co-workers.[18] These investigators recognized that although small peptides containing disulfide bonds could be reduced using cathodic reduction, there would likely be problems with proteins because of size and tertiary structure considerations. Therefore, a small thiol was used as a catalyst for the reduction.

Gorin and co-workers[19] have examined the rate of reaction of lysozyme with various thiols. At pH 10.0 (0.025 *M* borate), the relative rates of reaction were β-mercaptoethanol (2-mercaptoethanol), 0.2; dithiothreitol, 1.0; 3-mercaptopropionate, 0.4; and 2-amino-ethanethiol, 0.01. The results with 2-aminoethanethiol were somewhat surprising since the reaction (disulfide exchange) involves the thiolate anion and 2-aminoethanethiol would be more extensively ionized than the other mercaptans.

Finally, disulfide bonds can be cleaved by sulfite to from the *S*-sulfonate derivative as shown in Figure 3. The chemistry of this reaction has been reviewed by Cole.[20] The reaction proceeds optimally at alkaline pH (pH 9.0). It is necessary to include an oxidizing agent such as cupric ions, or as shown in Figure 3, *o*-iodosobenzoate to ensure effective conversion of all cystine residues to the corresponding *S*-sulfonate derivatives. The reaction is reversible to form cysteine upon treatment with a suitable mercaptan such as β-mercaptoethanol.

REFERENCES

1. **Anfinsen, C. B., Haber, E., Sela, M., and White, F. H., Jr.,** The kinetics of formation of native ribonuclease during oxidation of the reduced polypeptide chain, *Proc. Natl. Acad. Sci. U.S.A.*, 47, 1309, 1961.

2. **Givol, D., DeLorenzo, F., Goldberger, R. F., and Anfinsen, C. B.,** Disulfide interchange and the three-dimensional structure of proteins, *Proc. Natl. Acad. Sci. U.S.A.,* 53, 676, 1965.
3. **Hageman, T. C., Endres, G. F., and Scheraga, H. A.,** Mechanism of action of thrombin on fibrinogen. On the role of the A chain of bovine thrombin in specificity and in differentiating between thrombin and trypsin, *Arch. Biochem. Biophys.,* 171, 327, 1975.
4. **Cleland, W. W.,** Dithiothreitol, a new protective reagent for SH groups, *Biochemistry,* 3, 480, 1964.
5. **Iyer, K. S. and Klee, W. A.,** Direct spectrophotometric measurement of the rate of reduction of disulfide bonds. The reactivity of the disulfide bonds of bovine α-lactalbumin, *J. Biol. Chem.,* 248, 707, 1973.
6. **Gorecki, M. and Patchornik, A.,** Polymer-bound dihydrolipoic acid: a new insoluble reducing agent for disulfides, *Biochim. Biophys. Acta,* 303, 36, 1973.
7. **Fuchs, S., DeLorenzo, F., and Anfinsen, C. B.,** Studies on the mechanism of the enzymic catalysis of disulfide interchange in proteins, *J. Biol. Chem.,* 242, 398, 1967.
8. **Creighton, T. E., Hillson, D. A., and Freedman, R. B.,** Catalysis by protein-disulphide isomerase of the unfolding and refolding of proteins with disulphide bonds, *J. Mol. Biol.,* 142, 43, 1980.
9. **Janolino, V. G., Sliwkowski, M. Y., Swaisgood, H. F., and Horton, H. R.,** Catalytic effect of sulfhydryl oxidase on the formation of three-dimensional structure in chymotrypsinogen A., *Arch. Biochem. Biophys.,* 191, 269, 1978.
10. **Neumann, H. and Smith, R.L.,** Cleavage of the disulfide bonds of cystine and oxidized glutathione by phosphorothioate, *Arch. Biochem. Biophys.,* 122, 354, 1967.
11. **Neumann, H., Steinberg, J. Z., Brown, J. R., Goldberger, R. F., and Sela, M.,** On the non-essentiality of two specific disulphide bonds in ribonuclease for its biological activity, *Eur. J. Biochem.,* 3, 171, 1967.
12. **Light, A., Hardwick, B. C., Hatfield, L. M., and Sondack, D. L.,** Modification of a single disulfide bond in trypsinogen and the activation of the carboxymethyl derivative, *J. Biol. Chem.,* 244, 6289, 1969.
13. **Knights, R. J. and Light, A.,** Disulfide bond-modified trypsinogen. Role of disulfide 179-203 on the specificity characteristics of bovine trypsin toward synthetic substrates, *J. Biol. Chem.,* 251, 222, 1976.
14. **Mise, T. and Bahl, O. P.,** Assignment of disulfide bonds in the α-subunit of human chorionic gonadotropin, *J. Biol. Chem.,* 255, 8516, 1980.
15. **Gorin, G. and Godwin, W. E.,** The reaction of iodate with cystine and with insulin, *Biochem. Biophys. Res. Commun.,* 25, 227, 1966.
16. **Donovan, J. W.,** Spectrophotometric observation of the alkaline hydrolysis of protein disulfide bonds, *Biochem. Biophys. Res. Commun.,* 29, 734, 1967.
17. **Florence, T. M.,** Degradation of protein disulphide bonds in dilute alkali, *Biochem. J.,* 189, 507, 1980.
18. **Leach, S. J., Meschers, A., and Swanepoel, O. A.,** The electrolytic reduction of proteins, *Biochemistry,* 4, 23, 1965.
19. **Gorin, G., Fulford, R., and Deonier, R. C.,** Reaction of lysozyme with dithiothreitol and with other mercaptans, *Experientia,* 24, 26, 1968.
20. **Cole, R. D.,** Sulfitolysis, *Meth. Enzymol.,* 11, 206, 1967.

Chapter 8

THE MODIFICATION OF METHIONINE

The specific or selective modification of methionine (Figure 1) in proteins and peptides is somewhat difficult to achieve under relatively mild conditions. The majority of the modification reactions used to study methionine involve either oxidation or alkylation at the thioether sulfur.

Oxidation of methionine to methionine sulfoxide (Figure 2) can occur under a variety of conditions. Reagents for the "selective" oxidation of methionine which have attracted recent attention include chloramine T[1,2] (0.1 M phosphate, pH 7.0 or 0.1 M Tris, pH 8.4), sodium periodate[2] (0.1 M sodium acetate, pH 5.0), and hydrogen peroxide.[3] The reaction of methionine with chloramine T can be followed spectrophotometrically.[1] The reaction of chloramine T with methionine results in a significant change in the spectrum of chloramine T as shown in Figure 3. The use of this spectral change in the determination of methionine is shown in Figure 4. Cysteine interfered with this determination but other amino acids (i.e., tyrosine, tryptophan, histidine, serine) did not have any effect on the accuracy of analysis for methionine. It is noted that the oxidation of methionine is a possible side-reaction with the treatment of proteins with N-bromosuccinimide.[4]

It is possible to convert methionine sulfoxide to methionine under relatively mild conditions[5] thus providing for the reversibility of the oxidative reactions described above (Figure 5). This can be accomplished through both nonenzymatic and enzymatic methods. The nonenzymatic approaches have, in general, proved to be of greater value. A systematic study has shown that of four reducing agents tested, mercaptoacetic acid, β-mercaptoethanol, dithiothreitol, and N-methylmercaptoacetamide, the latter reagent, N-methylmercaptoacetamide was the most effective. The reactions demonstrated little pH dependence but did not proceed well at concentrations of acetic acid above 50% (v/v). Complete regeneration of methionine could be accomplished with 0.7 to 2.8 M reagent at 37° for 21 hr. An enzymatic system for the reduction of methionine sulfoxide has been reported.[6]

Methionine can be modified with various alkylation agents such as the α-halo acetic acids and their derivatives (Figure 6). The reaction of iodoacetate with methionine has been examined in some detail by Moore, Stein, and Gundlach.[7] The reaction of iodoacetate with methionine does not appear to be pH dependent and proceeds much slower than the reaction with cysteine under the mildly alkaline conditions used for reduction and carboxymethylation. The resulting sulfonium salt yields homoserine and homoserine lactone when heated at 100°C at pH 6.5. On acid hydrolysis (6 N HCl, 110°C, 22 hr), a mixture of methionine and S-carboxymethyl homocysteine together with a small amount of homoserine lactone was obtained. In general, methionine residues only react with the α-haloacids after the disruption of the secondary and tertiary structure of a protein.[8] Selectivity in the modification of methionine in proteins by α-halocids can be achieved by performing the reaction at acid pH (pH 3.0 or less). The modification of methionine by ethyleneimine has been reported in a reaction producing a sulfonium salt derivative.[9] The modification of methionine in azurin with bromoacetate has been reported.[10] In this protein, four of six methionine residues were modified at pH 4.0, while all methionine residues were reactive at pH 3.2. These modification reactions were performed in 0.1 M sodium formate at ambient temperature for 24 hr with 0.16 M bromoacetate. The modification of methionine in porcine kidney acyl CoA-dehydrogenase occurs with iodoacetate (0.030 M) in 0.1 M phosphate, pH 6.6 at ambient temperature.[11] The identification of methionine as the residue modified by iodoacetate in this protein was supported by the comparison of the chromatogram of the acid hydrolyzate of the modified protein (reacted with ^{14}C-iodoacetate) with that of the acid hydrolyzate of

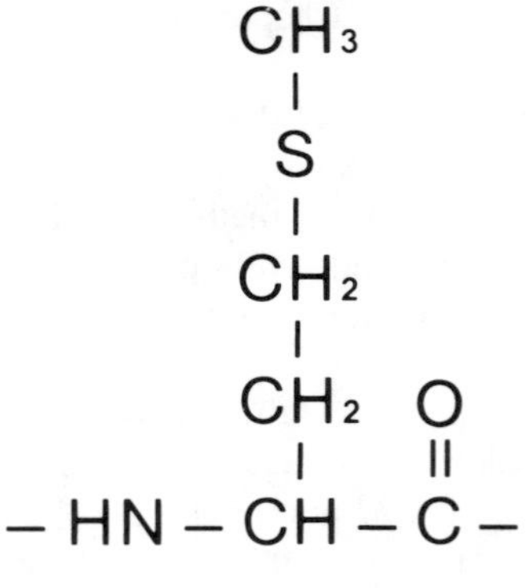

$$
\begin{array}{c}
\text{CH}_3 \\
| \\
\text{S} \\
| \\
\text{CH}_2 \\
| \\
\text{CH}_2 \quad\ \text{O} \\
| \qquad\ \| \\
-\,\text{HN}-\text{CH}-\text{C}-
\end{array}
$$

FIGURE 1.   The structure of methionine.

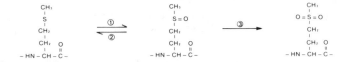

FIGURE 2.   The oxidation of methionine to methionine sulfoxide (step 1) which is reversible (step 2, see Figure 5), and subsequently to methionine sulfone.

synthetic *S*-([1-¹⁴C]carboxymethyl)-methionine.[7] This is necessary since the *S*-carboxymethyl derivative yielded several different compounds on acid hydrolysis.[7,12]

Naider and Bohak[13] have reported that the sulfonium salt derivatives of methionine (e.g., *S*-carboxymethyl methionine, the reaction product of methionine and iodoacetic acid) can be converted to methionine by reaction with a suitable nucleophile. For example, reaction of *S*-carboxamidomethyl methionine (in the peptide Gly-Met-Gly) with a sixfold molar excess of mercaptoethanol at pH 8.9 at a temperature of 30°C resulted in the complete regeneration of methionine after 24 hr of reaction. The *S*-phenacyl derivative of methionine (in the peptide Gly-Met-Gly) was converted to methionine in 1 hr under the same reaction conditions. These investigators also showed that chymotrypsin previously treated with phenacyl bromide under conditions which inactivate the enzyme concomitant with the alkylation of methionine-192[14] could be reactivated by treatment with β-mercaptoethanol at pH 7.5 (sodium phosphate). It is of interest that the *S*-phenacyl methionine in chymotrypsin is converted to methionine at a substantially faster rate than the tripeptide derivative. The authors speculate that the increased reactivity of the chymotrypsin derivative is a reflection of interaction of the phenacyl moiety with the substrate binding site.

Alkylation of methionyl residues in pituitary thyrotropin and lutropin with iodoacetic acid has been reported.[15] Differential reactivity of various methionyl residues was reported on reaction with iodoacetate in 0.2 *M* formate, pH 3.0 for 18 hr at 37°C.

Naider and co-workers[16] have examined the modification of a specific methionyl residue in β-galactosidase with *N*-bromoacetyl-β-D-galactosylamine. Reaction with *N*-bromoacetyl-β-D-galactosylamine in 0.1 *M* sodium phosphate, pH 7.5 (containing 1 m*M* MgCl₂) at 30°C led to loss of catalytic activity and modification of methionine. Reactivation of the modified enzyme with concomitant regeneration of methionine was achieved with β-mercaptoethanol (0.12 *M*) at pH 8.65 (90 min at 30°C).

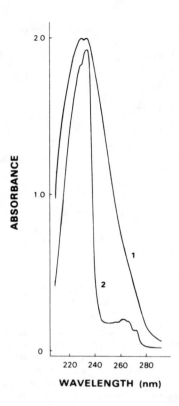

FIGURE 3. The UV absorption spectrum of chloramine-T (93.1 μg) in 3.0 mℓ 0.1 *M* sodium phosphaate, pH 7.0 before (curve 1) and after (curve 2) the addition of 99.4 μg methionine. The major change in the spectrum chloramine-T resulting from the interaction with methionine is a decrease in the width of the peak with λmax 234 nm resulting in a decrease in absorbance which is maximal between 244 and 248 nm. (From Trout, G. E., *Analyt. Biochem.*, 93, 419, 1979. With permission.)

MICROASSAY OF METHIONINE

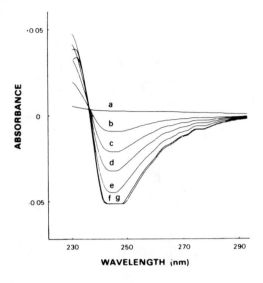

FIGURE 4.   The effect of increasing concentration of
chloramine-T on the difference spectrum resulting from
the interaction of chloramine-T with methionine as shown
in Figure 3. Curve a is the spectrum of 24.9 μg methionine in 3.0 mℓ of 0.1 *M* phosphate, pH 7.0. Spectra
b through g were obtained from the successive additions
of 10 μℓ portions of 10.0 m*M* chloramine-T. (From
Trout, G. E., *Analyt. Biochem.*, 93, 419, 1979. With
permission.)

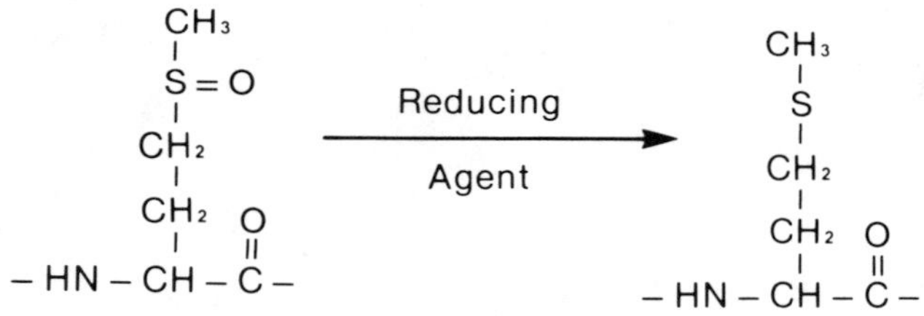

FIGURE 5.   The conversion of methionine sulfoxide to methionine in
the presence of a reducing agent.

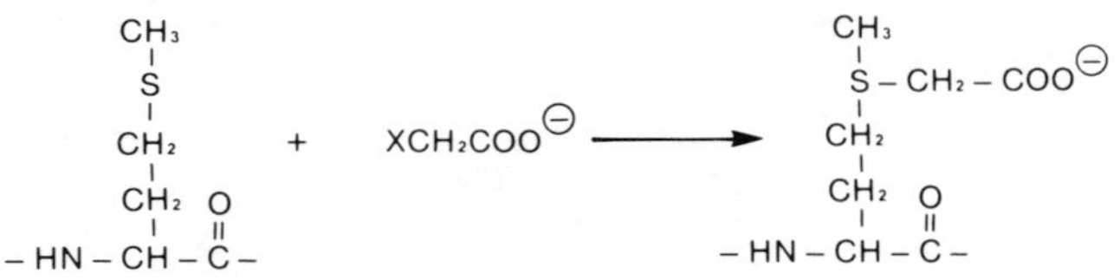

FIGURE 6.   The alkylation of methionine with an α-haloacid, such as
iodoacetic acid, to form the sulfonium salt derivative.

REFERENCES

1. **Trout, G. E.,** The estimation of microgram amounts of methionine by reaction with chloroamine-T, *Analyt. Biochem.,* 93, 419, 1979.
2. **de la Llosa, P., El Abed., A., and Roy, M.,** Oxidiation of methionine residues in lutropin, *Can. J. Biochem.,* 58, 745, 1980.
3. **Caldwell, P., Luk, D. C., Weissbach, H., and Brot, N.,** Oxidation of the methionine residues of *Escherichia coli* ribosomal protein L12 decreases the protein's biological activity, *Proc. Natl. Acad. Sci. U.S.A.,* 75, 5349, 1978.
4. **Spande, T. F. and Witkop, B.,** Determination of the tryptophan content of proteins with N-bromosuccinimide, *Meth. Enzymol.,* 11, 498, 1967.
5. **Houghten, R. A. and Li, C. H.,** Reduction of sulfoxides in peptides and proteins, *Analyt. Biochem.,* 98, 36, 1979.
6. **Brot, N., Weissbach, L., Werth, J., and Weissbach, H.,** Enzymatic reduction of protein-bound methionine sulfoxide, *Proc. Natl. Acad. Sci. U.S.A.,* 78, 2155, 1981.
7. **Gundlach, H. G., Moore, S., and Stein, W. H.,** The reaction of iodoacetate with methionine, *J. Biol. Chem.,* 234, 1761, 1959.
8. **Gurd, F. R. N.,** Carboxymethylation, *Meth. Enzymol.,* 11, 532, 1967.
9. **Schroeder, W. A., Shelton, J. R., and Robberson, B.,** Modification of methionyl residues during aminoethylation, *Biochim. Biophys. Acta,* 147, 590, 1967.
10. **Marks, R. H. L. and Miller, R. D.,** Chemical modification of methionine residues in azurin, *Biochem. Biophys. Res. Commun.,* 88, 661, 1979.
11. **Mizzer, J. P. and Thorpe, C.,** An essential methionine in pig kidney general acyl-CoA dehydrogenase, *Biochemistry,* 19, 5500, 1980.
12. **Goren, H. J., Glick, D. M., and Barnard, E. A.,** Analysis of carboxymethylated residues in proteins by an isotopic method and its application to the bromoacetate-ribonuclease reaction, *Arch. Biochem. Biophys.,* 126, 607, 1968.
13. **Naider, F. and Bohak, Z.,** Regeneration of methionyl residues from their sulfonium salts in peptides and proteins, *Biochemistry,* 11, 3208, 1972.
14. **Schramm, H. J. and Lawson, W. B.,** Über das activ Zentrum von Chymotrypsin. II. Modifizierung eines Methioninrestes in Chymotrypsin durch einfache Benzolderivate, *Hoppe-Seyler's Z. Physiol. Chem.,* 332, 97, 1963.
15. **Goverman, J. M. and Pierce, J. G.,** Differential effects of alkylation of methionine residues on the activities of pituitary thyrotropin and lutropin, *J. Biol. Chem.,* 256, 9431, 1981.
16. **Naider, F., Bohak, Z., and Yariv, J.,** Reversible alkylation of a methionyl residue near the active site of β-galactosidase, *Biochemistry,* 11, 3202, 1972.

Chapter 9

MODIFICATION OF HISTIDINE RESIDUES

The selective modification of histidyl residues in proteins had proved to be a difficult task until the development of diethylpyrocarbonate (ethoxyformic anhydride) which is described in considerable detail below. Until the development of this reagent, photo-oxidation had been the most generally used method for the modification of histidyl residues.

Photo-oxidation was used early for the modification of proteins,[1] but it was not until the work of Ray and Koshland,[2,3] that it proved possible to relate the modification of a specific residue to changes in biological activity. The technique of photo-oxidation has not proved of extensive value because of problems with the specificity of modification. Histidine, methionine, and tryptophan are quite sensitive to photo-oxidation while tyrosine, serine, and threonine are somewhat less sensitive.[2-4] Despite these difficulties, photo-oxidation is still used by investigators to explore the role of various functional groups in proteins and protein-protein complexes. Photo-oxidation was used to identify the protein(s) at the peptidyltransferase site of a bacterial ribosomal subunit.[5,6] Rose Bengal dye was used in these experiments. Histidine is the only amino acid modified under the reaction conditions. The time course for the loss of the various biological activities of reconstituted ribosomes on photo-oxidation is shown in Figure 1. Figure 2 shows the rate of histidine loss during photo-oxidation. The data are shown to be a summation of a minimum of three separate first-order reactions upon analysis as described by Ray and Koshland.[2] The rates of histidine loss were then compared to the rate of biological activity loss during photo-oxidation as shown in Figure 3. With the exception of EF-G·-GTP binding activity, the loss of biological activity (see Figure 1 for symbols) is most closely related to the "fast" histidine loss. In subsequent experiments, methylene blue dye (Eastman, dye content 91%) was used.[7] Peptidyltransferase activity was lost at a more rapid rate in the presence of methylene blue than Rose Bengal but data are not presented regarding any differences in residues modified or whether amino acid residues other than histidine are modified in the presence of methylene blue. Other investigators have also explored the effects of photo-oxidation on peptidyltransferase activity in *Escherichia coli* ribosomes.[8] These experiments were performed in 0.030 M Tris, 0.020 M $MgCl_2$, 0.220 KCl, pH 7.5 (9 mg ribosomes in 0.300 mℓ) with either eosin or Rose Bengal as the photo-oxidation agent. Irradiation was performed at 0 to 4°C using a 500 W slide projector (26 cm from condensor lens to sample) for 20 min. Photo-oxidation has also been used to study the role of histidine residues in polypeptide chain elongation factor Tu from *E. coli*.[9] The reaction is performed in 0.05 M Tris, 0.010 M Hg(OAc)$_2$, 0.005 M β-mercaptoethanol, 10% glycerol, pH 7.9. Irradiation is performed at 0 to 4°C with gentle stirring using a 375 W tungsten lamp at a distance of 15 cm. A glass plate was placed in the light beam to eliminate UV irradiation. The Rose Bengal dye is removed after 5 to 30 min from the reaction by chromatography on DEAE-Sephadex A-25 or A-50 equilibrated with 0.050 M Tris, pH 7.9, 0.010 M Mg (OAc)$_2$, 0.005 M β-mercaptoethanol, 10% glycerol. Amino acid analysis after acid hydrolysis (6 N HCl, 22 hr, 110°C, or 4 M methanesulfonic acid, 0.2% 2-aminoethylindal, 115°C, 24 hr for the determination of tryptophan) demonstrated that only histidine is modified (approximately 5/10 residues are modified; only one residue is modified in the presence of guanosine diphosphate).

Histidine residues can be modified by α-halo carboxylic acids and amides (i.e., iodoacetate and iodacetamide). In general, the histidine residue must have either enhanced nucleophilic character[10] or have been located in a unique microenvironment such as in ribonuclease.[11-14] Another approach to the modification of histidyl residues at the active site of certain enzymes (cf. serine proteases) has utilized peptide chloromethyl ketones[15-17] as affinity labels.

A compound related to α-halo carboxylic acids is *p*-bromophenacyl bromide, which has

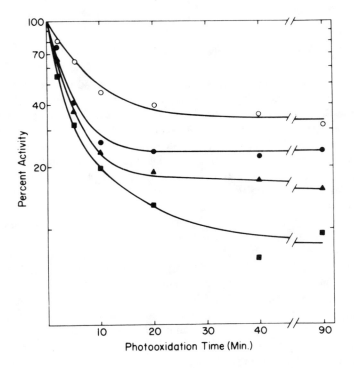

FIGURE 1.   The inactivation of biological activity of ribosome reconstituted with photo-oxidized ribosomal protein BL3. Ribosomal protein
BL3 was photo-oxidized as the B-13-23 S RNA complex in the presence
of rose bengal. Samples of the purified reconstituted ribosomes were assayed for EF-G·GTP binding (○), polyphenylalanine synthesis (●), peptidyltransferase activity (▲), and the ability to bind phenylalanyl-tRNA
(■). (From Auron, P. E., Erdelsky, K. J., and Fahnestock, S. R., *J. Biol.
Chem.*, 253, 6893, 1978. With permission.)

been demonstrated in several instances to modify histidyl residues in proteins. *p*-Bromo-
phenacyl bromide modifies a single histidine residue in taipoxin with a 350-fold decrease
in neurotoxicity.[18] The modification was performed in 0.1 *M* sodium cacodylate, pH 6.0,
0.1 *M* NaCl with an eightfold molar excess of *p*-bromophenacyl bromide* at 30°C for 22
hr. The reaction mixture was concentrated by lyophilization and subjected to gel filtration
(G-25 Sephadex in 0.1 *M* ammonium acetate) to remove excess reagent and buffer salts.
The protein fraction was taken to dryness as a salt-free preparation and subjected to a second
reaction with *p*-bromophenacyl bromide. The extent of modification was assessed by both
amino acid analysis (loss of histidine) and spectral analysis ($\Delta\epsilon$ 271 = 17,000 M^{-1} cm^{-1}).[19]
Two of seven histidine residues are modified (1 mol/mol in α-subunit; 1 mol/mol in β-
subunit) under these reaction conditions.

   The basic phospholipase A_2 from *Naja nigricollis* venom has been modified with *p*-
bromophenacyl bromide.[20] The modification was performed in 0.025 *M* Tris, pH 8.0 with
a tenfold molar excess of reagent at 30°C. After 40 min of reaction, the mixture was taken
to pH 4.0 with glacial acetic acid and taken through a G-25 Sephadex column. Amino acid
analysis after acid hydrolysis showed the loss of 1 mol of histidine/mole enzyme with no
other significant changes in composition. Subsequent analysis identified His[47] as the residue
modified.

---

*   It has been the authors' experience that this reagent is somewhat unstable and preparations *must* be recrystallized.

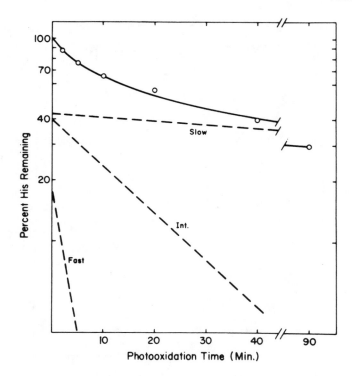

FIGURE 2. The loss of histidine residues occurring with photo-oxidation of Bl3 in the presence of rose bengal dye. The loss of histidine was assessed by amino acid analysis after acid hydrolysis. Loss of amino acids other than histidine was not observed under these experimental conditions. The solid line (○) shows the rate of loss of histidine residues. These data can be fit to represent three different classes of histidine residues as indicated by the broken lines. (From Auron, P. E., Erdelsky, K. J., and Fahnestock. S. R., *J. Biol. Chem.*, 253, 6893, 1978. With permission.)

The reaction of *p*-bromophenacyl bromide with pancreatic phospholipase A_2 has also been studied.[21] The reaction was performed in 0.1 *M* sodium cacodylate, pH 6.0.[22] Only histidine residues are modified under these conditions and it has been established that His^{48} is the residue modified. Under these reaction conditions the second order rate constant for the reaction of *p*-bromophenacyl bromide with porcine pancreatic phospholipase A_2 is 125 M^{-1} min^{-1} as compared to 79 M^{-1} min^{-1} for phenacyl bromide and 75 M^{-1} min^{-1} for 1-bromooctan-2-one. No reaction was observed with iodoacetamide under these reaction conditions. This same study[21] also explored the methylation of His^{48} at the N'-position on the imidazole ring with either methyl *p*-toluenesulfonate or methyl *p*-nitrobenzenesulfonate. Reaction with the latter reagent at pH 6.0 (0.050 *M* cacodylate) (40°C) is more rapid than with the former reagent.

Methyl *p*-nitrobenzenesulfonate has also been used to methylate histidine residue(s) in ribosomal peptidyl transferase.[23] In these experiments the ribosome preparation was modified by a 300-fold molar excess of methyl *p*-nitrobenzenesulfonate (from a stock solution dissolved in acetonitrile). The reaction took place in 0.01 *M* Tris, pH 7.4, 0.008 *M* $MgCl_2$, 0.05 *M* Nh_4Cl, 1 μM puromycin at 24°C for 45 min. The author suggests only histidine residues are modified but definitive evidence on this point is absent.

As indicated in the brief discussion regarding the modification of histidyl residue with α-haloalkyl carboxylic acids, inherent nucleophilicity and environmental considerations are of considerable importance.

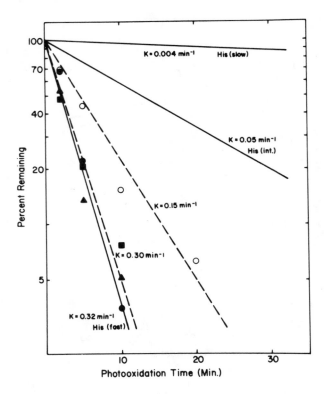

FIGURE 3.   A comparison of the rates of the loss of activity and histidine as a function of time of photooxidation. The solid lines represent the first-order rates for the loss of histidine replotted as a percentage of each class of reactive histidyl residues (see Figure 2). The broken line indicates the loss of activity (for symbols see Figure 1 caption). (From Auron, P. E., Erdelsky, K. J., and Fahnestock, S. R., *J. Biol. Chem.*, 253, 6893, 1978. With permission.)

Another example of the modification of histidine by reagents which, in general, react more avidly with other residues is the reaction of D-amino acid oxidase with dansyl chloride.[24] In this study, D-amino acid oxidase was allowed to react with a fivefold molar excess of dansyl chloride (from a stock solution in acetone; final concentration of acetone in the reaction mixture did not exceed 5% final volume) in 0.05 M phosphate, pH 6.6. The reaction was terminated by the addition of benzoate, insoluble material removed by centrifugation, and the mixture passed through a G-25 column equilibrated with 0.06 M phosphate, 0.010 M benzoate, pH 6.6. Reaction with dansyl chloride under these conditions resulted in virtually complete inactivation of the enzyme with the incorporation of 1.7 mol of reagent/mole enzyme. Substantially complete reactivation occurred with 0.5 M hydroxylamine (NH$_2$OH) at pH 6.6. This reactivation excluded reaction with primary amino functional groups such as lysine, and amino acid analysis suggested the reaction had not occurred with an oxygen nucleophile such as tyrosine. Treatment of the enzyme with diethylpyrocarbonate also resulted in the loss of catalytic activity and reduced the amount of dansyl groups incorporated in a subsequent reaction suggesting that dansyl chloride reacts with the same functional group that reacted with diethylpyrocarbonate.

As mentioned above, diethylpyrocarbonate (ethoxyformic anhydride) has become useful during the past 5 to 10 years for studies involving the specific modification of histidine. In the pH range from 5.5 to 7.5, diethylpyrocarbonate is reasonably specific for reaction with histidyl residues. There are several studies of the reaction under more acidic conditions.[25,26] In one of the studies, polymerization of ribonuclease was observed both in deionized water (presumably at acidic pH) and in 0.1 M Tris, pH 7.2. Maleylation of ribonuclease obviated polymerization, suggesting that the amino groups were involved in this cross-linking reaction. Work with diethylpyrocarbonate through 1975 has been reviewed by Miles.[27]

Reaction of diethylpyrocarbonate with histidine residues at a *moderate excess* of diethyl-pyrocarbonate results in substitution at one of the nitrogen positions on the imidazole ring. This reaction is associated with an increase in absorbance at 240 nm ($\Delta\epsilon = 3200\ M^{-1}$ cm^{-1}). The modification is readily reversed at alkaline pH and, in particular, in the presence of nucleophiles such as Tris. Generally treatment with neutral hydroxylamine (0.1 to 1.0 M, pH 7.0) is used to regenerate histidine. As with the deacylation of O-acetyl tyrosine by neutral hydroxylamine (see Volume II, Chapter 3), the higher the concentration of hydroxylamine, the more rapid the process of decarboxyethylation. Disubstitution on the imidazole ring, carboxyethylation at both the N_1 and N_3 positions, results in a derivative with altered spectral properties compared to the monosubstituted derivative. This derivative does not regenerate histidine and treatment with neutral hydroxylamine or base results in scission of the imidazole ring. With disubstitution, a loss of histidine is detected by amino acid analysis after acid hydrolysis. The monosubstituted derivative is unstable under conditions of acid hydrolysis and yields free histidine. Reaction can also occur with other nucleophiles such as cysteine, tyrosine, and primary amino groups. Modification at sulfhydryl residues, not well documented with protein-bound cysteine, can be determined by a decrease in free sulfhydryl groups. Reaction of tyrosine is easily assessed by a decrease in absorbance at 275 to 280 nm similar to that observed on O-acetylation with N-acetylimidazole (see Volume II, Chapter 3). This modification is reversed by neutral hydroxylamine. Reaction at primary amino groups (α-amino groups; ϵ-amino groups of lysine) results in a derivative which is stable to hydroxylamine.

As described in some detail by Miles,[27] the reagent is very sensitive to base-catalyzed hydrolysis. At ambient temperature, the $T_{1/2}$ for the hydrolysis of diethylpyrocarbonate at pH 7.0 (phosphate) is less than 10 min and is markedly shorter with increasing pH. Increasing the pH not only decreases reagent stability (and thus the concentration of one component of a second order reaction over the time period studied) but also increases the possibility of reaction at primary amine functional groups. In our laboratory we have found it convenient to use dilute (0.025 to 0.100 M) phosphate buffer, pH 6.0 for our studies. We prepare stock solutions of diethylpyrocarbonate in *anhydrous* ethanol. These solutions are used within a few hours and the actual concentration of reagent obtained by the stoichiometry of reaction with imidazole in the pH 6.0 buffer using the increase in absorbance at 230 nm to monitor the reaction ($\Delta\epsilon = 3 \times 10^3\ M^{-1}\ cm^{-1}$)[25,27] both before and after a given series of experiments. It should be noted that there is not complete agreement regarding the magnitude of the spectral change as a result of carboxyethylation. The value $\Delta\epsilon = 3200\ M^{-1}\ cm^{-1}$ at 242 nm is given above. Other investigators have used the value of $\Delta\epsilon = 3600\ M^{-1}\ cm^{-1}$ at 240 nm.[28,29] A value of $\Delta\epsilon = 3500\ M^{-1}\ cm^{-1}$ at 242 nm has also been reported.[30] Most of these values have been obtained by the use of reaction with model compounds such as N-acetylhistidine. Neurath and co-workers[30] have noted that the use of spectral analysis of histidine modification with intact protein is greatly complicated if modification of tyrosine is occuring concomitantly. This situation results from the marked decrease in absorbance of tyrosine at 234 nm which will decrease the magnitude of the net absorbance increase in the 230 to 250 nm range resulting from histidine modification. Roosemont[31] has considered this problem in some detail. The greater the excess of diethylpyrocarbonate used the less reliable the value for $\Delta\epsilon$ obtained with *known* stoichiometric modification of model histidine (imidazole derivative) compounds. This is shown in Figure 4 where increasing the ratio of diethylpyrocarbonate to histidine results in species with increased absorbance. It is suggested

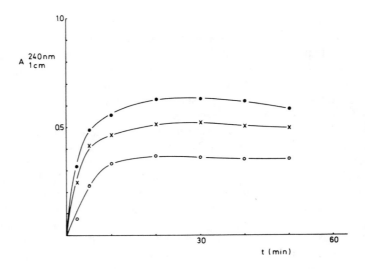

FIGURE 4. The time course for the increase in absorbance at 240 nm occurring upon reaction between histidine and diethylpyrocarbonate (DEP) as a function of diethylpyrocarbonate concentration. The reaction mixtures were 0.1 mM with respect to histidine in 0.1 M acetate, pH 6.0 at 2.3 mM DEP (○), 11.5 mM DEP(x) and 23 mM DEP (●). (From Roosemont, J. L., *Anal. Biochem.*, 88, 314, 1978. With permission.)

that the commonly used $\Delta\epsilon = 3,200\ M^{-1}\ cm^{-1}$ can only be used at low concentrations of diethylpyrocarbonate.

Selected examples of the use of diethylpyrocarbonate to study the function of histidyl residues in proteins are presented in Table 1.

Certain of these studies need to be considered in further detail. Neurath and co-workers[30] have examined the reaction of thermolysin with diethylpyrocarbonate (ethoxyformic anhydride) in some detail. The time course for the loss of catalytic activity by thermolysin upon reaction with diethylpyrocarbonate is shown in Figure 5. Also shown in Figure 5 is the recovery of activity upon reaction of the modified protein with hydroxylamine. Under these reaction conditions 13.4 carboethoxy groups were incorporated per mole of enzyme. The presence of carbobenzoxy-L-phenylalanine, a competitive inhibitor, protected the enzyme from inactivation and reduced the extent of modification to 12.5 mol/mol protein. It should be noted that the reaction demonstrated a pH dependence (Figure 6) consistent with the modification of histidine. Figure 7 shows the difference spectra of thermolysin (panel A) after reaction with diethylpyrocarbonate (solid line) and then after subsequent reaction with 0.020 M neutral hydroxylamine (broken line). This is compared with panel B where the difference spectrum obtained on the reaction of diethylpyrocarbonate with N-acetyl-L-tyrosine ethyl ester (solid line) or imidazole (broken line); the dotted line is the algebraic sum of the two reactions. The increase at 242 nm is consistent with the modification of histidine while the decrease at 278 to 280 nm is indicative of tyrosine modification. It is of interest that the tyrosyl residues modified during the reaction were *not* regenerated on reaction with 0.020 M hydroxylamine (Figure 8). These investigators also make the point that there is a significant decrease in the absorbance of N-acetyl-L-tyrosine ethyl ester at 234 nm (Figure 7, panel B) which would affect the accuracy of measurement at 240 to 242 nm for the determination of the extent of histidine modification. These investigators found it more accurate to determine the extent of histidine modification by spectral measurement during hydroxylamine reactivation.

McCormick and co-workers[32] have examined the reaction of diethylpyrocarbonate with pyridoxamine (pyridoxine)-5'-phosphate oxidase. The modification reaction was performed

Table 1
THE REACTION OF DIETHYLPYROCARBONATE WITH HISTIDYL RESIDUES IN PROTEINS

Protein	Solvent/temp.	Reagent excess[a]	Extent of modification[b]	Second order rate constant[c] ($\ell\ M^{-1}\ min^{-1}$)	Other amino acid modified	Ref.
Thermolysin	0.05 M CaCl$_2$, 0.1 M NaCl, 0.025 M HEPES, pH 7.2/25°C	—	2/$_d$	—	Tyr (2), Lys (8)	1
Themolysin	0.05 M CaCl$_2$ 0.025 M HEPES, pH 5.7/25°C	—	1/^{d}	—	Tyr (1), Lys (2)	1
Bacterial Luciferase	0.1 M phosphate/pH 6.1/0°C	600	3/	146[e]		2
Crotoxin	Half-saturated NaOAc	1,000		—	—	3
Prostatic acid phosphatase	0.025 M sodium barbital, 0.15 M NaCl, pH 6.9/25°C	5,000	12/27	7[e]	f	
Pyridoxamine-5′-phosphate oxidase	0.1 M phosphate, pH 7.0/25°C	500	4/7	750[g]	Cys (1/6)[h]	5
Yeast enolase	0.050 M ADA, 0.001 M MgCl$_2$,. 0.01 mM EDTA, pH 6.1/0°C	1,390	6/^{i}	55.0[j] 10.7[t]	k	6
Fructose bisphosphatase	0.050 M acetate, pH 6.5	4,000 8,000 40,000	5.3/13 8.8/13 13/13[l]	—	l	7
Ribulose bisphosphate carboxylase	0.050 or Tris, 0.001 M EDTA, 0.020 M MgCl$_2$ pH 7.0/30°C	—	2/2	m	n	8
Eschen chia coli elongation factor Tu	—[o]	—	—	—	—	10
L-α-Hydroxy acid oxidase	0.020 M MES, pH 7.0/25°C	—	2/[p]	690[q]	r	11
Thiamine binding protein from *Saccaromyces cerevisiae*	0.050 M sodium phosphate, pH 7.0/25°C	—	—	120.5[s]	—	12
Clostridium histolyticum collagenase	0.050 M HEPES, 0.010 M CaCl$_2$, pH 17.5/22°C	—	—	—	—	13
Lactate dehydrogenase	0.1 M sodium pyrophosphate, pH 7.2/10°C	—	—	10380 522	—	14
Mitochondrial nicotinamide nucleotide transhydrogenase	0.025 M sucrose, 0.020 M MES, pH 5.9/23°C	—	—	—	—	15
Alcohol oxidase	0.050 M sodium$_v$ phosphate, pH 7.5/0°C	—	4/	25.2	w	16
D-β-hydroxybutyrate dehydrogenase	0.020 MES, 5 M rotenone, pH 6.0/	—	—	—	—	8

Table 1 (continued)
THE REACTION OF DIETHYLPYROCARBONATE WITH HISTIDYL RESIDUES IN PROTEINS

Protein	Solvent/temp.	Reagent excess[a]	Extent of modification[b]	Second order rate constant[c] ($\ell\ M^{-1}\ min^{-1}$)	Other amino acid modified	Ref.
Benzodiazepine receptor	0.01 M sodium phosphate, 0.2 M NaCl, pH 6.0	—	—	—	—	19
Scrapie agent	0.020 M Tris, pH 7.4 containing 1 mM EDTA and 0.2% Sarkosyl/ 23°C	—	—	—	x	20
Succinate dehydrogenase	0.24 M sucrose, 0.100 M potassium phosphate, pH 6.0/ 0°C	—	—	—	y	21
Ribulose *bis*-phosphate carboxylase/ oxygenase	0.050 M Tris, 0.001 M EDTA, 0.020 M MgCl$_2$, pH 7.0/21°C	—	3.4/[a]	2340	—	22
Transferrin	0.01 M potassium phosphate, 0.05 KCl, pH 6.1	—	54—72%[aa]	—	—	23
Dihydrofolate reductase	0.05 M Tris, pH 7.5/10°C	—	6/7	29	bb	24

[a]   Moles/mole of protein.

[b]   Residues His modified/total His in protein.

[c]   For reaction with histidine.

[d]   Inactivation was demonstrated to result from the modification of a single histidine residue.

[e]   Assuming loss of activity is a direct indication of a single histidine modification.

[f]   There was only partial recovery of activity upon treatment with hydroxylamine (0.2 M, pH 7.0, 25°C). Two residues of histidine were lost as assessed by amino acid analysis after acid hydrolysis without loss of other amino acids suggesting that disubstitution has occured on the imidazole ring of certain histidine residues.

[g]   For inactivation of catalytic activity. A value of 51.6 $M^{-1}\ sec^{-1}$ (3096 $M^{-1}\ min^{-1}$) was calculated for the pH-independent second order rate constant.

[h]   No direct determination of primary amino modification is reported. Activity is recovered by neutral hydroxylamine (0.9 M). Direct determination of tryptophan and tyrosine revealed no loss of these residues.

[i]   Obtained from reaction at either 0 or 25°C.

[j]   Data obtained at pH 6.6 (0.050 M N- (2-acetamido) iminodiacetic acid, ADA), at 0°C. Two phase reaction was observed.

[k]   Spectral analysis did not indicate tyrosine modification. Possible primary amine modification was not determined. The loss in catalytic activity was reversed by 0.25 M hydroxylamine, pH 7.0.

[l]   Addition of more diethylpyrocarbonate results in further increases in absorbance at 242 nm suggesting disubstitution on the imidazole ring of histidyl residues. Spectral analysis did not suggest modification of tyrosine under the reaction conditions. Possible modification of primary groups was not assessed.

[m]   The reaction of diethylpyrocarbonate with ribulose bisphosphate carboxylase shows saturation kinetics (k = 7.3 mM) suggesting "specific" binding of diethylpyrocarbonate to the enzyme prior to the reaction resulting in inactivation. Data are not presented to show a similar phenomena with the actual reaction of histidyl residues in the enzyme.

[n]   Activity was recovered by treatment with hydroxylamine (0.4 M NH$_2$OH, pH 7.0, 48 hr at 4°C increased activity from 55 to 89%; similar treatment at 25°C resulted in similar activity recovery in 1 hr) The authors note that reaction of diethylpyrocarbonate with cysteine (*N*-acetylcysteine) also results in an increase in absorbance at 240 nm that is reversed by hydroxylamine. This reaction apparently occurs only in carboxylate buffers (e.g. acetate or succinate) and has been noted by other investigators.[9] The reaction product of diethylpyrocarbonate with cysteine is considerably less stable than *N*-carbethoxyimidazole derivatives.

Table 1 (continued)

ᵒ The crystalline enzyme preparation (7 nmoles) [washed with 41% $(NH_4)_2 SO_4$] was dissolved in 0.600 mℓ, 0.010 M Tris, pH 7.0 containing 5 mM $MgCl_2$, 0.100 M KCl and 10 μM guanosine diphosphate. The pH of this solution was then adjusted to 6.0 with 1.0 M sodium cacodylate — 0.050 M $MgCl_2$.

ᵖ Per FMN (hence per dimer, therefore this would be 4 residues/tetramer).

ᑫ Determined from rate of loss of catalytic activity.

ʳ No reaction at cysteine, tryptophan or tyrosine is observed under these reaction conditions.

ˢ Determined from rate of loss of thiamine binding activity.

ᵗ Thiomethylated at cysteine-165 (reaction with methyl methanethiosulfonate). Enzyme remains catalytically active but with reduced affinity for pyruvate and lactate.

ᵘ The second order rate constant for the native enzyme is 10,920 M^{-1} min^{-};1. These values were obtained from the measurement of the rate of loss of enzyme activity. Virtually identical values were obtained from direct measurement of histidine modification by spectroscopy.

ᵛ Diethylpyrocarbonate introduced as acetonitrile solution.

ʷ Incorporation of radiolabeled diethylpyrocarbonate was closely related to extent of histidine modification as assessed by spectroscopy ($\Delta\epsilon = 3900$ M^{-1} cm^{-1} for monosubstituted derivative). Hydroxylamine treatment did not result in recovery of enzyme activity although radiolabel was lost. The enzyme was inactivated by 10 mM hydroxylamine at neutral pH, 0°C. This is not an infrequent observation from our consideration of the literature in this area. Although very few investigators (we have not found any report) have examined the possibility of peptide bond cleavage with hydroxylamine at neutral pH, the possibility cannot be disregarded considering the cleavage of Asn-Gly bonds under more alkaline conditions.[17]

ˣ Inactivation reversed by 0.100 to 0.5 M hydroxylamine. The pH of this reaction is not specified.

ʸ Submitochondrial particles were used in this study. The inactivation produced by diethylpyrocarbonate is partially reversed by neutral (pH 7.0) hydroxylamine. The extent of activity recovery was dependent on hydroxylamine with maximum activity recovery at 0.020 M hydroxylamine decreasing significantly at 0.115 M hydroxylamine.

ᶻ Stoichiometry determined by spectral analysis ($\Delta\epsilon = 3200$ M^{-1} cm^{-1} at 240 nm) (3.4 residues modified) is in excellent agreement with amount of radiolabeled diethylpyrocarbonate incorporated (3.5).

ᵃᵃ Varied with species source of transferrin: human 14/7; rabbit, 14/18; human lactotransferrin, 7/10; bovine lactotransferrin, 7/9; chicken ovotransferrin, 9/14.

ᵇᵇ There is no reaction with tyrosine under these conditions. Reaction at primary amine functions was not excluded. Only partial reactivation is obtained upon treatment with hydroxylamine (approximately 50% recovery with 1.0 M hydroxylamine; no reaction at 0.1 M hydroxylamine).

References for Table 1

1. **Burstein, Y., Walsh, K. A., and Neurath, H.,** Evidence of an essential histidine residue in thermolysin, *Biochemistry,* 13, 205, 1974.
2. **Cousineau, J. and Meighen, E.,** Chemical modification of bacterial luciferase with ethoxyformic anhydride: evidence for an essential histidyl residue, *Biochemistry,* 15, 4992, 1976.
3. **Jeng, T.-W. and Fraenkel-Conrat, H.,** Chemical modification of histidine and lysine residues of crotoxin, *FEBS Lett.,* 87, 291, 1978.
4. **McTigue, J. J. and van Etten, R. L.,** An essential active-site histidine residue in human prostatic acid phosphatase. Ethoxyformylation by diethyl pyrocarbonate and phosphorylation by a substrate, *Biochim. Biophys. Acta,* 523, 407, 1978.
5. **Horiike, K., Tsuge, H., and McCormick, D. B.,** Evidence for an essential histidyl residue at the active site of pyridoxamine (pyridoxine)-5′-phosphate oxidase from rabbit liver, *J. Biol. Chem.,* 254, 6638, 1979.
6. **George, A. L., Jr. and Borders, C. L., Jr.,** Chemical modification of histidyl and lysyl residues in yeast enolase, *Biochim. Biophys. Acta,* 569, 63, 1979.
7. **Demaine, M. M. and Benkovic, S. J.,** Selective modification of rabbit liver fructose bisphosphatase, *Arch. Biochem. Biophys.,* 205, 308, 1980.
8. **Saluja, A. K. and McFadden, B. A.,** Modification of histidine at the active site of spinach ribulose bisphosphate carboxylase, *Biochem. Biophys. Res. Commun.,* 94, 1091, 1980.
9. **Garrison, C. K. and Himes, R. H.,** The reaction between diethylpyrocarbonate and sulfhydryl groups in carboxylate buffers, *Biochem. Biophys. Res. Commun.,* 67, 1251, 1975.
10. **Jonák, J. and Rychik, I.,** Chemical evidence for the involvement of histidyl residues in the functioning of *Escherichia coli* elongation factor Tu, *FEBS Lett.,* 117, 167, 1980.
11. **Meyer, S. E. and Cromartie, T. H.,** Role of essential histidine residues in L-α-hydroxy acid oxidase from rat kidney, *Biochemistry,* 19, 1874, 1980.

12. **Nishimura, H., Sempuku, K., and Iwashima, A.,** Possible functional roles of carboxyl and histidine residues in a soluble thiamine-binding protein of *Saccharomyces cerevisiae, Biochim. Biophys. Acta,* 668, 333, 1981.

13. **Bond, M. D., Steinbrink, D. R., and Van Wart, H. E.,** Identification of essential amino acid residues in *Clostridium histolyticum* collagenase using chemical modification reactions, *Biochem. Biophys. Res. Commun.,* 102, 243, 1981.

14. **Bloxham, D. P.,** The chemical reactivity of the histidine-195 residue in lactate dehydrogenase thiomethylated at the cysteine-165 residue, *Biochem. J.,* 193, 93, 1981.

15. **Phelps, D. C. and Hatefi, Y.,** Inhibition of the mitochondrial nicotinamide nucleotide transhydrgenase by dicyclohexylcarbodiimide and diethylpyrocarbonate, *J. Biol. Chem.,* 256, 8217, 1981.

16. **Cromartie, T. H.,** Sulfhydryl and histidinyl residues in the flavoenzyme alcohol oxidase from *Candida boidinii, Biochemistry,* 20, 5416, 1981.

17. **Bornstein, P. and Balian, G.,** Cleavage at Asn-Gly bonds with hydroxylamine, *Meth. Enzymol.,* 47, 132, 1977.

18. **Phelps, D. C. and Hatefi, Y.,** Inhibition of D-β-hydroxybutyrate dehydrogenase by butanedione, phenylglyoxal and diethyl pyrocarbonate, *Biochemistry,* 20, 459, 1981.

19. **Burch, T. P. and Ticku, M. K.,** Histidine modification with diethylpyrocarbonate shows heterogeneity of benzodiazepine receptors, *Proc. Natl. Acad. Sci. U.S.A.,* 78, 3945, 1981.

20. **McKinley, M. P., Masiarz, F. R., and Prusiner, S. B.,** Reversible chemical modification of the scrapie agent, *Science,* 214, 1259, 1981.

21. **Vik, S. B. and Hatefi, Y.,** Possible occurrence and role of an essential histidyl residue in succinate dehydrogenase, *Proc. Natl. Acad. Sci. U.S.A.,* 78, 6749, 1981.

22. **Saluja, A. K. and McFadden, B. A.,** Modification of the active site histidine in ribulosebisphosphate carboxylase/oxygenase, *Biochemistry,* 21, 89, 1982.

23. **Mazurier, J., Leger, D., Tordera, V., Montreuil, J., and Spik, G.,** Comparative study of the iron-binding properties of transferrins. Differences in the involvement of histidine residues as revealed by carbethoxylation, *Europ. J. Biochem.,* 119, 537, 1981.

24. **Daron, H. H. and Aull, J. L.,** Inactivation of dihydrofolate reductase from *Lactobacillus casei* by diethyl pyrocarbonate, *Biochemistry,* 21, 737, 1982.

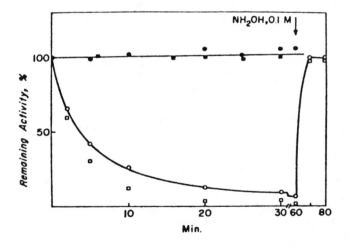

FIGURE 5.   The inactivation of thermolysin (15 μ*M*) by diethylpyrocarbonate (3 m*M* in 0.020 *M*.HEPES, 0.005 *M* CaCl₂, pH 7.2 in the presence (●,■) or absence (○,□) of 5 m*M* Cbz-phenylalanine. Catalytic activity was determined with either furylacryloylglycyl-L-leucinamide (●,○) or casein (■,□). The arrow indicates the addition of neutral hydroxylamine (pH 7.2) to a final concentration of 0.1 *M*. (From Burstein, Y., Walsh, K. A., and Neurath, H., *Biochemistry,* 13, 205, 1974. With permission.)

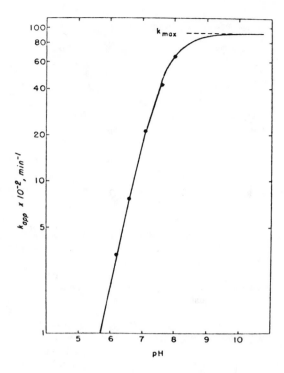

FIGURE 6.   The pH dependence for the pseudo first-order rate constant for the inactivation of thermolysin by diethylpyrocarbonate. The line is a theoretical curve calculated from

$$k_{app} = \frac{k_{max}}{1 + [H^+]K_a}$$

assuming a single ionizing group of pKa = 7.6 and k_{max} = 0.91 min^{-1}. (From Burstein, Y., Walsh, K. A., and Neurath, H., *Biochemistry,* 13, 205, 1974. With permission.)

at pH 7.0 (0.1 *M* potassium phosphate containing 5% (v/v) EtOH) at 25°C generally in the presence of flavin mononucleotide (FMN). Figure 9 shows the time course for the modification of the oxidase under these experimental conditions. Also shown in Figure 9 is the dependence of the observed first-order rate constant on diethylpyrocarbonate concentration (panel B). The inset plot shows the reaction to be second-order with a rate constant of 12.5 M^{-1} sec^{-1}. Figure 10 shows the pH dependence for the inactivation process. The loss of catalytic activity appeared to be correlated with the modification of one of the four histidyl residues in this protein. This residue appears to be more accessible for modification as shown in Figure 11.

There are several other studies which provide insight into the chemistry of the reaction of diethylpyrocarbonate with proteins. Saluja and McFadden[33] have explored the reaction of diethylpyrocarbonate with spinach ribulose bisphosphate carboxylase. One interesting observation is that the plot of half-inactivation time vs. the reciprocal of diethylpyrocarbonate concentration suggested that saturation kinetics existed consistent with the ''affinity'' binding of reagent prior to protein modification. Of greater general value is the study of the spectral changes occuring during the modification of this enzyme as shown in Figure 12. The solid line shows the change in absorbance at 242 nm of the carboxylase upon reaction with

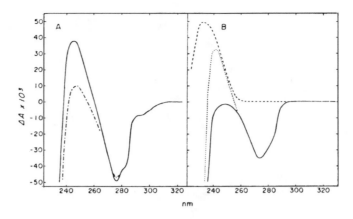

FIGURE 7.   (A) Difference UV absorption spectra of thermolysin after inactivation by diethylpyrocarbonate and reactivation by neutral hydroxylamine. Equal volumes (2.5 mℓ of enzyme (15 μM) in 0.02 M HEPES, 0.1 NaCl, 5 mM CaCl₂) were placed in two spectrophotometer cuvettes. Diethypyrocarbonate (25 μℓ of 0.3 M solution in ethanol) was added to one cuvette and an equal volume of ethanol added to the other cuvette. The difference spectra were determined after 40 min (solid line). Hydroxylamine (25 μℓ of a 2.0 M solution, pH 7.0) was then added to both cells and the difference spectra determined 40 min later (broken line). (B) Spectral changes of model compounds after treatment with diethylpyrocarbonate (15 μM) in 0.025 M HEPES, pH 7.0. The solid line is 50 μM *N*- acetyl-L-tyrosine ethyl ester while the broken line is 2 mM imidazole. The dotted line is the algebraic sum of the two difference spectra. (From Burstein, Y., Walsh, K. A., and Neurath, H., *Biochemistry,* 13, 205, 1974. With permission.)

diethylpyrocarbonate (the magnitude of the increase is consistent with the modification of 2.4 histidine residues per combination of small subunit and large subunit). This change is completely reversed on treatment with hydroxylamine. Also shown is the change in absorbance at 242 nm on the reaction of *N*-acetylcysteine with diethylpyrocarbonate in 0.1 *M succinate* pH 6.4. This spectral change occurs only in *carboxylate* buffers and is comparatively transient when compared to the protein reaction product. The study of Bloxham[34] on the reactivity of the active site histidine in lactate dehydrogenase is particularly fascinating. The rate of reaction of the histidine residue in the native enzyme was compared to the thiomethyl derivative (prepared by reaction with methyl methanethiosulfonate) as shown in Figure 13. There is a substantial decrease in the nucleophilic character of the active site histitine (histidine-195). Cromartie[35] has examined the modification of alcohol oxidase with diethylpyrocarbonate in 0.050 *M* sodium phosphate, pH 7.5, at 0°C, The UV difference spectrum of the enzyme before (a) and after (b) the addition of diethylpyrocarbonate is shown in Figure 14. No evidence for tyrosine modification is seen under these reaction conditions. Further kinetic analysis supported the modification of histidine as the event responsible or inactivation of catalytic activity. It is, however, of interest to note that treatment with neutral hydroxylamine (0.010 to 0.100 *M*) did not result in the recovery of catalytic activity although most of the radiolabeled reagent was removed ([1-¹⁴C]-diethylpyrocarbonate). It was observed that 0.010 *M* hydroxylamine caused an 80% loss of alcohol oxidase activity at 0°C. As mentioned above, reaction of diethylpyrocarbonate at pH values above 7.0 does increase the possibility of amino group modification. This is demonstrated by the observation of Van Wart and co-workers[36] on the reaction of diethylpyrocarbonate with a bacterial collagenase. As shown in Figure 15, reaction of the enzyme with diethyl-

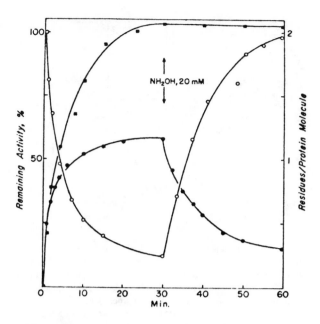

FIGURE 8.   The results of the modification of thermolysin with diethylpyrocarbonate (ethoxyformic anhydride). Thermolysin (15 μM) was allowed to react with DEP (3 mM) in 0.025 M HEPES, pH 7.2 followed by reaction with hydroxylamine (20 mM) indicated by the arrow. The open circles show the changes in catalytic activity. Modification of tyrosyl residues (■) was monitored by the decrease in absorbance at 278 nm using a $\Delta\epsilon$ = 1310 M^{-1} cm^{-1}. Modification of histidyl residues was followed by the increase in absorbance at 242 nm (●) using a $\Delta\epsilon$ = 3200 M^{-1} cm^{-1}. (From Burstein, Y., Walsh, K. A., and Neurath, H., *Biochemistry*, 13, 205, 1974. With permission.)

pyrocarbonate results in the loss of catalytic activity. Reaction with hydroxylamine did not markedly restore enzymatic activity. Hydroxylamine by itself did not have a deleterious effect on catalytic activity as treatment with this reagent did restore activity lost on the modification of tyrosyl residues with *N*-acetylimidazole. These investigators consider modification of the ϵ-amino group(s) of lysine with diethylpyrocarbonate to be a more likely cause for the irreversible lost of enzymic activity. Daron and Aull[37] have studied the reaction of diethyl pyrocarbonate with dihydrofolate reductase *(Lactobacillus casei)*. The UV spectra of the enzyme before and after reaction with diethylpyrocarbonate are shown in Figure 16. Catalytic activity is lost on reaction with diethylpyrocarbonate but is partially recovered on reaction with 1.0 m hydroxylamine (pH 7.5) but not with 0.1 M hydroxylamine (pH 7.5.) (Figure 17). Clearly the reaction of proteins with hydroxylamine after modification with diethylpyrocarbonate must be carefully studied in order to obtain meaningful results.

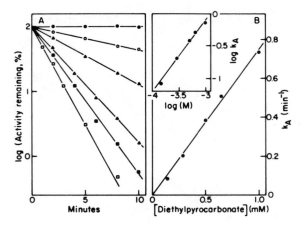

FIGURE 9.   The inactivation of pyridoxamine-5'-phosphate oxidase by diethylpyrocarbonate. Panel A describes an experiment
where the enzyme (2.32 μM) was incubated in 0.1 M potassium
phosphate, pH 7.0 containing 3.75μM riboflavin 5'-phosphate in
the absence (●) or presence of diethylpyrocarbonate at the following concentrations 0.144 mM (○), 0.287 mM (▲), 0.501 mM
(△), 0.644 mM (■), or 1.00 mM (□). Panel B shows the dependence of the pseudo first-order rate constants for inactivation
on the concentration of diethylpyrocarbonate. Values for k_A were
determined from the slopes of the semilogarithmic plots of panel
A. The inset in panel B shows a plot of k_A vs. log concentration
of diethylpyrocarbonate. (From Horiike, K., Tsuge, H., and
McCormick, D. B., *J. Biol. Chem.*, 254, 6638, 1979. With
permission.)

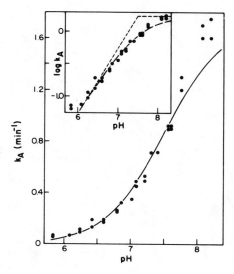

FIGURE 10. The dependence of the rate constant (k_A; see Figure 9) for the inactivation of pyridoxamine-5′-phosphate oxidase by diethylpyrocarbonate. The pseudo first-order rate constants (k_A) were calculated from the slopes of the semilogarithmic plots of activity vs. time (see Figure 9, panel A). The solid curve is drawn assuming there was no reaction between the protonated residue (pKa = 7.5). The second-order rate constant for the reaction of diethylpyrocarbonate with the unprotonated enzyme was 51.6 M^{-1} sec^{-1}. The inset shows a plot of k_A vs. pH. (From Horiike, K., Tsuge, H., and McCormick, D. B., *J. Biol Chem.*, 254, 6638, 1979. With permission.)

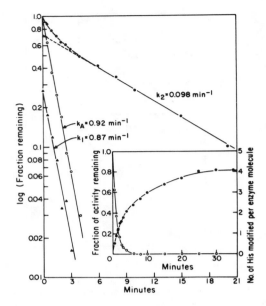

FIGURE 11.   The relationship between the extent of histidine modification and the loss of catalytic activity of pyridoximine-5′-phosphate oxidase upon reaction with diethylpyrocarbonate. The holoenzyme (2.4 μ*M*) was incubated with 1.2 m*M* diethylpyrocarbonate in 0.1 *M* potassium phosphate, pH 7.0. At the indicated times, the number of *N*-carboethoxyhistidyl residues was determined by spectrophotometric analysis and the fraction of remaining histidine residues (●) calculated by taking the total number of such modifiable residues (4.1) as unity and enzyme activity (○) determined. The fast phase of the modification (▲) was obtained by subtracting the contribution of the slow phase (dashed line) and replotting the differences. The olid curveis calculated on the basis of the following equation

$$x = (n - m)/n = 0.27e^{-0.87t} + 0.73e^{-0.098t}$$

where x is the total fraction of residues remaining after reaction, n is the total number of modifiable residues, m is the number of residues actually modified, and t is time. The inset shows a plot of activity remaining (○) and the number of histidyl residues modified per enzyme molecule (●) as a function of reaction time. (From Horiike, K., Tsuge, H., and McCormick, D. B., *J. Biol. Chem.*, 254, 6638, 1979. With permission.)

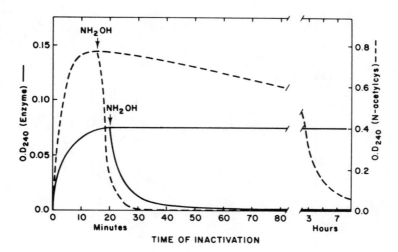

FIGURE 12.. The reaction of spinach ribulose bisphosphate carboxylase with diethylpyrocarbonate in 0.1 *M* phosphate, pH 6.4 (solid line). The dashed line shows the reaction of diethylpyrocarbonate with *N*-acetylcysteine in 0.1 *M* succinate buffer, pH 6.4. The arrow indicates the point of addition of hydroxylamine, pH 7.0, to a final concentration of 0.4 *M*. A difference spectrum of the modified vs. unmodified enzyme (not shown) demonstrated an absorption maximum at 240 nm. (From Saluja, A. K. and McFadden, B. A., *Biochem. Biophys. Res. Commun.*, 94, 1091, 1980. With permission.)

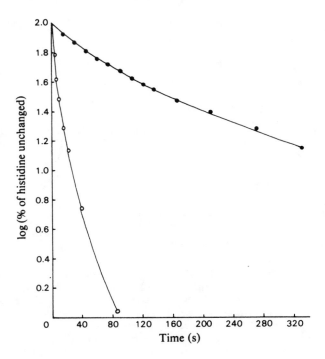

FIGURE 13.   A comparison of the reaction rate of native (○) and thiomethylated (●) lactate dehydrogenase with diethylpyrocarbonate. The thiomethylated enzyme was prepared by reaction with methyl methanethiosulfonate. The proteins were in 0.1 *M* sodium phosphate, pH 7.2 at 10°C and allowed to react with 1 m*M* diethylpyrocarbonate. The extent of histidine modification was assessed by the increase in absorbance at 240 nm. ($\Delta\epsilon = 3600 \ M^{-1} \ cm^{-1}$.) (From Bloxham, D. P., *Biochem. J.*, 193, 93, 1981. With permission.)

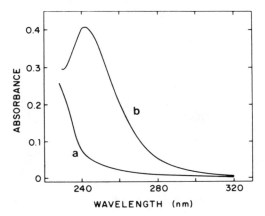

FIGURE 14.   Ultraviolet difference spectrum for the reaction of *Candida boidinii* flavoenzyme alcohol oxidase with diethylpyrocarbonate in 0.05 *M* sodium phosphate, pH 7.5. Curve a was obtained before the addition of diethylpyrocarbonate while curve b was obtained after 30 min of reaction of protein in the sample cuvette with diethylpyrocarbonate. (From Cromartie, T. H., *Biochemistry*, 20, 5416, 1981. With permission.)

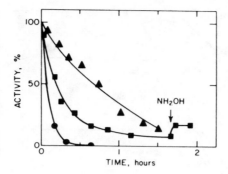

FIGURE 15. A time course study of the inactivation of *Clostridium histolyticum* collagenase by diethylpyrocarbonate. The reaction was performed in 0.05 M HEPES, 10 mM CaCl$_2$ with 30 mM diethylpyrocarbonate (●) or 10 mM diethylpyrocarbonate in the presence (▲) or absence (■) of carbobenzoxy-Gly-Pro-Gly-Gly-Pro-Ala. The arrow indicates the point of addition of hydroxylamine to a final concentration of 0.5 M. (From Bond, M. D., Steinbrink, D. R., and Van Wart, H. E., *Biochem. Biophys. Res. Commun.*, 102, 243, 1981. With permission.)

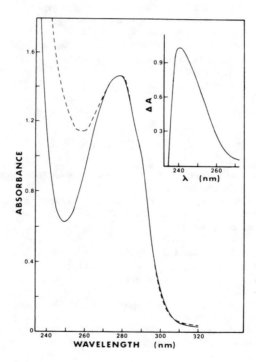

FIGURE 16. The UV absorption spectra of dihydrofolate reductase in the presence or absence of diethylpyrocarbonate. The reactions were performed in 0.05 M Tris-HCl, pH 7.5, in the presence (dashed line) or absence (solid line) of 3.24 mM diethylpyrocarbonate. (From Daron, H. H. and Aull, J. L., *Biochemistry*, 21, 737, 1982. With permission.)

124   *Chemical Reagents for Protein Modification*

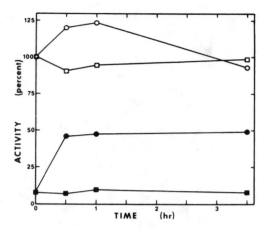

FIGURE 17.   Reaction of diethylpyrocarbonate-inac-
tivated dihydrofolate reductase by hydroxylamine. The
dihydrofolate reductase was incubated either in the presence (solid symbols) or absence (closed symbols) of 6.9
m*M* diethylpyrocarbonate. The diethylpyrocarbonate-
modified enzyme had 8% of the activity of the control
enzyme preparation. The reaction mixtures were then
treated with hydroxylamine (pH 7.5) at a final concentration of either 1.0 *M* (circles) or 0.1 *M* squares). (From
Daron, H. H., and Aull, J. L., *Biochemistry*, 21, 737,
1982. With permission.)

REFERENCES

1. **Weil, L., James, S., and Buchert, A. R.,** Photo-oxidation of crystalline chymotrypsin in the presence of methylene blue, *Arch. Biochem. Biophys.*, 46, 266, 1953.
2. **Ray, W. J., Jr., and Koshland, D. E., Jr.,** A method for characterizing the type and numbers of groups involved in enzyme action, *J. Biol. Chem.*, 236, 1973, 1961.
3. **Ray, W. J., Jr., and Koshland, D. E.,** Identification of amino acids involved in phosphoglucomutase action, *J. Biol. Chem.*, 237, 2493, 1962.
4. **Bond, J. S., Francis, S. H., and Park, J. H.,** An essential histidine in the catalytic activities of 3-phosphoglyceraldehyde dehydrogenase, *J. Biol. Chem.*, 245, 1041, 1970.
5. **Fahnestock, S. R.,** Evidence of the involvement of a 50S ribosomal protein in several active sites, *Biochemistry*, 14, 5321, 1975.
6. **Auron, P. E., Erdelsky, K. J., and Fahnestock, S. R.,** Chemical modification studies of a protein at the peptidyltransferase site of the *Bacillus stearothermophilus* ribosome. The 50S ribosomal subunit is a highly integrated functional unit, *J. Biol. Chem.*, 253, 6893, 1978.
7. **Dohme, F. and Fahnestock, S. R.,** Identification of proteins involved in the peptidyl transferase activity of ribosomes by chemical modification, *J. Mol. Biol.*, 129, 63, 1979.
8. **Cerna, J. and Rychlik, I.,** Photoinactivation of peptidyl transferase binding sites, *FEBS Lett.*, 102, 277, 1979.
9. **Nakamura, S. and Kaziro, Y.,** Selective photooxidation of histidine residues in polypeptide chain elongation factor Tu from *E. coli, J. Biochem. (Tokyo)*, 90, 1117, 1981.
10. **Inagami, T. and Hatano, H.,** Effect of alkylguanidines on the inactivation of trypsin by alkylation and phosphorylation, *J. Biol. Chem.*, 244, 1176, 1969.
11. **Stark, G. R., Stein, W. H., and Moore, S.,** Relationships between the conformation of ribonuclease and its reactivity toward iodoacetate, *J. Biol. Chem.*, 236, 436, 1961.
12. **Heinrikson, R. L., Stein, W. H., Crestfield, A. M., and Moore, S.,** The reactivities of the histidine residues at the active site of ribonuclease toward halo acids of different structures, *J. Biol. Chem.*, 240, 2921, 1965.

13. **Fruchter, R. G., and Crestfield, A. M.,** The specific alkylation by iodoacetamide of histidine 12 in the active site of ribonuclease, *J. Biol. Chem.,* 242, 5807, 1967.

14. **Lin, M. C., Stein, W. H., and Moore, S.,** Further studies on the alkylation of the histidine residues of pancreatic ribonuclease, *J. Biol. Chem.,* 243, 6167, 1968.

15. **Schoellmann, G. and Shaw. E.,** Direct evidence for the presence of histidine in the active center of chymotrypsin, *Biochemistry,* 2, 252, 1963.

16. **Shaw, E., Mares-Guia, M., and Cohen, W.,** Evidence for an active center histidine in trypsin through use of a specific reagent, 1-chloro-3-tosylamido-7-amino-2-heptanone, the chloromethyl ketone derived from N^α-tosyl-L-lysine, *Biochemistry,* 4, 2219, 1965.

17. **Segal, D. M., Powers, J. C., Cohen, G. H., Davies, D. R., and Wilcox, P. E.,** Substrate binding site in bovine chymotrypsin Aγ. A crystallographic study using peptide chloromethyl ketones as site-specific inhibitors, *Biochemistry,* 10, 3728, 1971.

18. **Fohlman, J., Eaker, D., Dowdall, M. J., Lüllmann-Rauch, R., Sjödin, T., and Leander, S.,** Chemical modification of taipoxin and the consequences for phospholipase activity, pathophysiology, and inhibition of high-affinity choline uptake, *Eur. J. Biochem.,* 94, 531, 1979.

19. **Halpert, J., Eaker, D., and Karlsson, E.,** The role of phospholipase activity in the action of a presynaptic neurotoxin of *Notechis scutatus scutatus* (Australian Tiger Snake), *FEBS Lett.,* 61, 72, 1976.

20. **Yang, C. C. and King, K.,** Chemical modification of the histidine residue in basic phospholipase A$_2$ from the venom of *Naja nigricollis, Biochim. Biophys. Acta,* 614, 373, 1980.

21. **Verheij, H. M., Volwerk, J. J., Jansen, E. H. J. M., Puyk, W. C., Dijkstra, B. W., Drenth, J., and de Haas, G. H.,** Methylation of histidine-48 in pancreatic phospholipase A$_2$. Role of histidine and calcium ion in the catalytic mechanism, *Biochemistry,* 19, 743, 1980.

22. **Volwerk, J. J., Pieterson, W. A., and de Haas, G. H.,** Histidine at the active site of phospholipase A$_2$, *Biochemistry,* 13, 1446, 1974.

23. **Glick, B. R.,** The chemical modification of *Escherichia coli* ribosomes with methyl *p*-nitrobenzenesulfonate. Evidence for the involvement of a histidine residue in the functioning of the ribosomal peptidyl transferase, *Can. J. Biochem.,* 58, 1345, 1980.

24. **Nishino, T., Massey, V., and Williams, C. H., Jr.,** Chemical modifications of D-amino acid oxidase. Evidence for active site histidine, tyrosine, and arginine residues, *J. Biol. Chem.,* 255, 3610, 1980.

25. **Melchior, W. B. Jr. and Fahrney, D.,** Ethoxyformylation of proteins. Reaction of ethoxyformic anhydride with α-chymotrypsin, pepsin and pancreatic ribonuclease at pH 4, *Biochemistry,* 9, 251, 1970.

26. **Wolf, B., Lesnaw, J. A., and Reichmann, M. E.,** A mechanism of the irreversible inactivation of bovine pancreatic ribonuclease by diethylpyrocarbonate. A general reaction of diethylpyrocarbonate with proteins, *Eur. J. Biochem.,* 13, 519, 1970.

27. **Miles, E. W.,** Modification of histidyl residues in proteins by diethylpyrocarbonate, *Meth. Enzymol.,* 47, 431, 1977.

28. **Holbrook, J. J. and Ingram, V. A.,** Ionic properties of an essential histidine residue in pig heart lactate dehydrogenase, *Biochem. J.,* 131, 729, 1973.

29. **Cousineau, J. and Meighen, E.,** Chemical modification of bacterial luciferase with ethoxyformic anhydride: evidence for an essential histidyl residue, *Biochemistry,* 15, 4992, 1976.

30. **Burstein, Y., Walsh, K. A., and Neurath, H.,** Evidence of an essential histidine residue in thermolysin, *Biochemistry,* 13, 205, 1974.

31. **Roosemont, J. L.,** Reaction of histidine residues in proteins with diethylpyrocarbonate: differential molar absorptivities and reactivities, *Anal. Biochem.,* 88, 314, 1978.

32. **Horiike, K., Tsuge, H., and McCormick, D. B.,** Evidence for an essential histidyl residue at the active site of pyridoxamine (pyridoxine)-5'-phosphate oxidase from rabbit liver, *J. Biol. Chem.,* 254, 6638, 1979.

33. **Saluja, A. K. and McFadden, B. A.,** Modification of histidine at the active site of spinach ribulose bisphosphate carboxylase, *Biochem. Biophys. Res. Commun.,* 94, 1091, 1980.

34. **Bloxham, D. P.,** The chemical reactivity of the histidine-195 residue in lactate dehydrogenase thiomethylated at the cysteine-165 residue, *Biochem. J.,* 193, 93, 1981.

35. **Cromartie, T. H.,** Sulfhydryl and histidinyl residues in the flavoenzyme alcohol oxidase from *Candida boidinii, Biochemistry,* 20, 5416, 1981.

36. **Bond, M. D., Steinbrink, D. R., and Van Wart, H. E.,** Identification of essential amino acid residues in *Clostridium histolyticum* collagenase using chemical modification reactions, *Biochem. Biophys. Res. Commun.,* 102, 243, 1981.

37. **Daron, H. H. and Aull, J. L.,** Inactivation of dihydrofolate reductase from *Lactobacillus casei* by diethyl pyrocarbonate, *Biochemistry,* 21, 737, 1982.

Chapter 10

THE MODIFICATION OF LYSINE

The chemical modification of lysine residues in proteins is based upon the ability of the ϵ-amino group of this residue to react as a nucleophile. Under normal reaction conditions, lysyl residues are the second strongest nucleophiles in a protein molecule; cysteine is the most reactive nucleophile. However, for lysine to function optimally as a nucleophile, the proton usually bound to lysyl residues at physiological pH must be removed. This is shown in Figure 1.

The protonated form is essentially unreactive. The pK_a of an ''average'' lysyl residue in a protein is 10 (see Table 2 in Chapter 6). The majority of modification reactions are performed at pH 8.0 to 9.0.

It is somewhat difficult to selectively modify lysyl residues in proteins. A number of the reagents which are used to modify lysyl residues also have the potential to react with the N-terminal amino group(s), with tyrosyl residues and with cysteinyl residues.

Lysine residues can be modified by reaction with α-ketoalkyl halides such as iodoacetic acid. This reaction has been discussed in detail by Gurd.[1] Alkylation can occur at pH greater than 7.0 but the rate of reaction is much slower than reaction with cysteinyl residues. Both the mono- and disubstituted derivatives have been reported. The monosubstituted derivative migrates close to methionine on amino acid analysis while the disubstituted derivative migrates near aspartic acid. It should be noted that reaction with α-keto alkyl halides is not considered particularly useful for the modification of primary amino groups. This reaction can be a possible side reaction occurring during the reduction and carboxymethylation of proteins. The reactivity of a given lysyl residue is affected by the nature of surrounding amino residues.

The reaction of lysyl residues with aryl halides has been of considerable value. Both fluoronitrobenzene and fluorodinitrobenzene have been of considerable value in protein chemistry since Sanger and Tuppy's work on the structure of insulin.[2] Carty and Hirs developed the use of 4-sulfonyl-2-nitrofluorobenzene for the modification of amino groups in pancreatic ribonuclease with 4-sulfonyloxy-2-nitrofluorobenzene as a function of pH. This is a particularly useful experiment since it is critical to understand that these investigators actually measured the amount of native protein remaining by chromatographic fractionation. As would be expected, the rate of modification increases with increasing pH. This reagent also is more stable than, for example, fluorodinitrobenzene under alkaline conditions permitting more accurate measurement at pH greater than 9.0. The lysine residue at position 41 is the site of major substitution which is a reflection of the lower pKa for the ϵ-amino group of this residue. Use of this compound did not present the solubility and reactivity problems posed by the fluoronitrobenzene compounds. It was possible to qualitatively determine the classes of amino groups in ribonuclease; these were the α-amino group, nine ''normal'' amino groups and lysine 41. The reactivity of lysine 41 was influenced by neighboring functional groups. This effect was lost at pH greater than 11 or on thermal denaturation of the protein. The reaction of 1-dimethylaminonaphthalene-5-sulfonyl chloride (dansyl chloride) has been useful both in the structural analysis and amino group modification with proteins. In one study,[4] dansyl chloride (in acetone) is added to a solution of trypsin in 0.1 M phosphate, pH 8.0. The reaction is terminated after 24 hr at 25°C by acidification to pH 3.0 with 1.0 M HCl. Insoluble material is removed by centrifugation and the supernatant fraction placed in dialysis. These investigators reported modification of the amino-terminal isoleucine and one lysine residue. The extent of modification was determined by absorbance at 336 nm ($\epsilon_m = 3.4 \times 10^4\ M^{-1}\ cm^{-1}$).

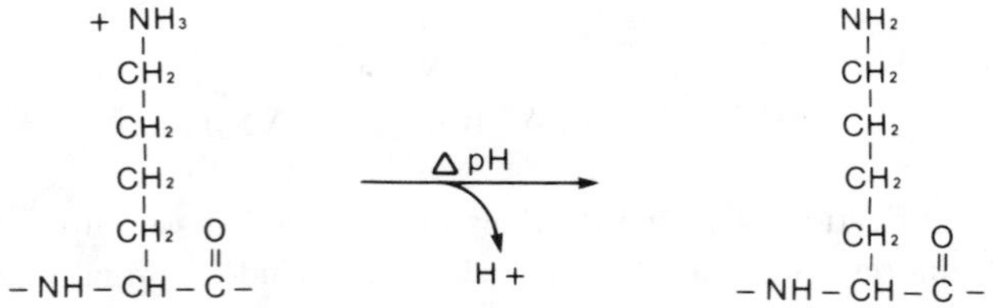

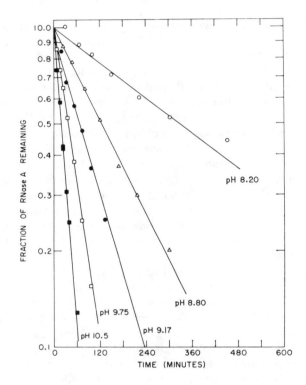

FIGURE 1.   The structure of lysine.

FIGURE 2.   The reaction of bovine pancreatic ribonuclease A with 4-sulfonyloxy-2-nitrofluorobenzene (potassium salt) as a function of pH. The pH was maintained by addition of 0.2 N NaOH during the course of the reaction at 28°C. The amount of ribonuclease remaining was determined by chromatographic analysis (Amerlite® IRC-50). (From Carty, R. P. and Hirs, C. H. W., *J. Biol. Chem.*, 243, 5254, 1968. With permission.)

The reaction of 2-carboxy-4,6-dinitrochlorobenzene with proteins has been explored.[5] This reagent reacts with amino, sulfhydryl and amino groups. At pH 8.2, the reaction with sulfhydryl groups is much more rapid than with lysine (for cysteine, 0.0560 M^{-1} sec^{-1}, for lysine 0.0031 M^{-1} sec^{-1}; for proline, 0.0001 M^{-1} sec^{-1}. The extent of modification can be determined from spectral analysis of the product, which has an absorption maximum at 370 nm ($\Delta\epsilon = 1.58 \times 10^3$ M^{-1} sec^{-1}). The UV spectra for ribonuclease and ribonuclease modified with 2-carboxy-4,6-dinitrochlorobenzene (CDNP-RNAse) are shown in Figure 3. It is possible to separate the modified protein from the unmodified protein as shown in Figure 4. The modification of pancreatic ribonuclease is performed at pH 8.2 (pH-stat) for

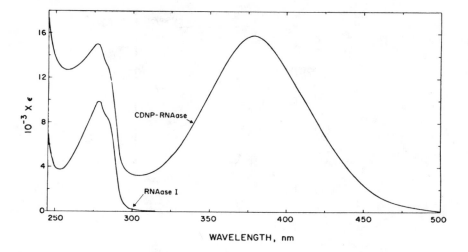

FIGURE 3.   Ultraviolet spectroscopy of the product of the reaction of 2-carboxy-4,6-dinitro-chlorobenzene with bovine pancreatic ribunoclease (CDNP-RNAase) at pH 8.2 (pH-stat). The modified protein was separated by chromatography prior to spectral analysis (Bio-Rex 70; see Figure 4). Shown for comparison is the spectrum of the unmodified enzyme. The spectra were determined in 0.1 *M* sodium phosphate, pH 7.5. (From Bello, J., Iijima, H., and Kartha, G., *Int. J. Peptide Protein Res.,*14, 199, 1979. With permission.)

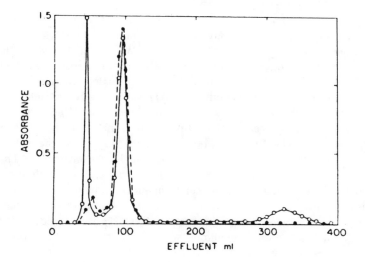

FIGURE 4.   The purification of CDNP-RNAase on a 0.9 × 110 cm column of Bio-Rex 70 (100—200 mesh). The solvent used was 0.2 *M* sodium phosphate, pH 6.1. Absorbance was monitored at 280 nm (○) and 370 nm (●). The product (CDNP-RNAase) eluted at 100 mℓ while unmodified RNAase eluted at 330 mℓ and reagent at 45 mℓ. (From Bello, J., Iijima, H., and Kartha, G., *Int. J. Peptide Protein Res.*, 14, 199, 1979. With permission.)

2 hr at ambient temperature. Hiratsuka and Uchida[6] examined the reaction of *N*-methyl-2-anilino-6-naphthalenesulfonyl chloride with lysyl residues in cardiac myosin. There was a difference in the nature of the reaction in the presence and absence of divalent cation. *N*-methyl-2-anilino-6-naphthalenesulfonyl chloride has been suggested for use as a fluorescent probe for hydrophobic regions of protein molecules.[7,8] The extent of incorporation of the

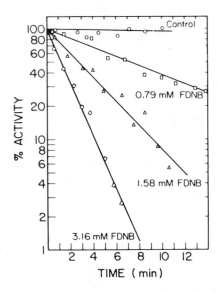

FIGURE 5.   The inactivation of luciferase by 2,4-dinitrofluorobenzene. The reactions were performed in 0.05 M sodium phosphate, pH 7.0 at 25°C and were initiated by the addition of 2,4-dinitrofluorobenzene (FDNB) in ethanol. Shown are the control (○), 0.79 mM FDNB (□), 1.58 mM FDNB (△) and 3.16 mM FDNB(◇).

N-methyl-2-anilino-6-naphthalenesulfonyl moeity into protein can be determined by spectral analysis at 327 nm ($\Delta\epsilon = 2.0 \times 10^4 \ M^{-1} \ cm^{-1}$).[6,7] Welches and Baldwin[9] have recently examined the reaction of bacterial luciferase with 2,4-dinitrofluorobenzene. The fluorescence of N-methyl-2-anilino-6-naphthalenesulfonyl derivatives is extremely sensitive to the polarity of the medium.[8] Reaction occurred at both cysteinyl residues and lysyl residues. Modification was associated with inactivation at the rate of 157 $M^{-1} \ min^{-1}$ at pH 7.0 (0.05 M phosphate). The loss of activity was not reversed by treatment with β-mercaptoethanol under conditions in which the free sulfhydryl group should be regenerated by thiolysis. The reactive sulfhydryl group could be reversibly blocked with methyl methanethiolsulfonate and then treated with 2,4-dinitrofluorobenzene. The time course for the *inactivation* of luciferase is shown in Figure 5. Both lysyl and cysteinyl residues can be modified under the experimental conditions (0.05 M phosphate, pH 7.0 at 25°C) used in these studies. In order to assess the significance of reaction at primary amino groups, the cysteinyl residues were ''blocked'' with methyl methanethiolsulfonate. Reaction of luciferase with methyl methanethiolsulfonate resulted in greater than 95% loss of catalytic activity (twofold molar excess of methyl methanethiolsulfonate in 0.02 M phosphate, pH 7.0 at 25°C). The loss of activity can be completely reversed with β-mercaptoethanol (97 mM). The small amount of residual activity present after treatment with methyl methanethiolsulfonate is further reduced on treatment with 2,4-dinitrofluorobenzene and the recovery of activity subsequent to β-mercaptoethanol is greatly reduced (see Figure 6). Quantitative analysis was not performed but qualitative analysis suggested that the modification occurred at the α-amino group of methionine or the α- and/ or β-subunits. The effects of pH on the reaction of fluorodinitrobenzene with luciferase is shown in Figure 7. It is of interest to compare the rate of reaction of fluorodinitrobenzene with model compounds and luciferase as has been done by these investigators as shown in Figure 8. Note that the rate of reaction with luciferase is much faster than with any of the model compounds.

Acylation of amino groups in proteins by reaction with carboxylic acid anhydrides has been extensively used. Riordan and Vallee[10] have discussed the process of acetylation in

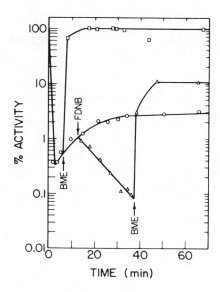

FIGURE 6. Protection of the thiol functional group reactivity with 2,4-dinitrofluorobenzene (FDNB) by prior reaction with methyl methanethiolsulfonate. Three portions of the methyl methanethiolsulfonate-luciferase were studied: the control preparation (○), a second portion treated with 97 mM β-mercaptoethanol (BME) (□) (indicated by the BME arrow), and a third portion (△) which was allowed to reaction with FDNB (indicated by the FDNB arrow) which was allowed to proceed until 90% of the activity had been lost at which point β-mercaptoethanol (97 mM) was added (indicated by the BME arrow at approximately 40 min). (From Welches, W. R. and Baldwin, T. O., *Biochemistry,* 20, 512, 1981. With permission.)

some detail. Acetylation is generally carried out with acetic anhydride at alkaline pH in either a pH-stat or in saturated sodium acetate. Performing the modification reaction under these latter conditions (saturated sodium acetate) results in increased specificity since *O*-acetyl tyrosine is unstable in sodium acetate. Acetylation has been used to study calcitonin[11] and a bacterial cytochrome.[12] Acetic and maleic anhydride have been used to study elastase.[13] In these studies, the reaction was carried out in a pH-stat to maintain alkaline pH. Reaction occurred at both lysyl and tyrosyl residues. It is relatively easy to differentiate between the two sites of modification since *O*-acyl tyrosyl residues are unstabe at pH ≥ 9.0. Studies with maleic anhydride showed that the amino terminal valine was not available for modification at pH 8.0 to 9.0 but could be modified at pH 11.0. Modification of this residue could be achieved in the presence of urea at a lower pH.

More recently, trifluoroacetylated derivatives have been of interest in the study of protein structure. In these studies, ethylthiotrifluoroacetate was used to modify cytochrome C in 0.14 *M* sodium phosphate, pH 8.0.[14] The pH was maintained at pH 8.0 using a pH-stat. Singly-substituted derivatives of cytochrome C can be separated by chromatography on anion-exchange resin (Bio Rex 70) and carboxymethylcellulose. It is critical to avoid lyophilization during the preparation of the various derivatives. These derivatives have been subjected to further investigation[15,16] including the use of ^{19}F-containing derivatives for nuclear magnetic resonance probes.[17] Radiolabeled acetic anhydride has been used in competitive labeling experiments to determine the reactivity of individual residues.[18,19]

Succinic anhydride has also proved useful in the modification of lysine.[20] Modification of lysine residues with succinic anhydride results in charge reversal. Reaction with succinic anhydride frequently results in the dissociation of multimeric proteins and has also been

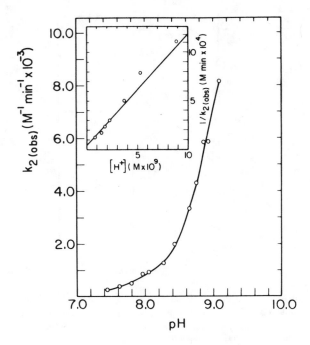

FIGURE 7. The pH dependence for the reaction of FDNB with luciferase. The observed second-order rate constant is plotted as a function of pH (0.05 *M* pyrophosphate). The inset shows a plot of the reciprocal of the observed second-order rate constant as a function of hydrogen ion concentration permitting the evaluation of the absolute second-order rate constant for the reaction (k_2) ($2.4 \times 10^5 \ M^{-1} \ min^{-1}$) and an apparent pKa of 9.4. (From Welches, W. R. and Baldwin, T. O., *Biochemistry*, 20, 512, 1981. With permission.)

reagent or protein	functional group	$\bar{k}_{2(obsd)}$ ($M^{-1} \cdot min^{-1}$)
2-mercaptoethanol[a]	SH	5.5
N-acetylcysteine[a]	SH	6.3
N^{α}-acetyllysine[a]	NH$_2$	0.013
methionine[a]	NH$_2$	0.070
luciferase[b]	NH$_2$	157

[a] Reactions were performed at pH 7.0, 20 °C, in 0.02 M phosphate buffer; the progress of the reaction was monitored spectrophotometrically. [b] The apparent second-order rate constant for reaction with luciferase was derived from the data in Figure 1.

FIGURE 8. The apparent second-order rate constants for the reaction of various functional groups with 2,4-dinitrofluorobenzene at pH 7.0 (0.02 *M* phosphate, 20°C). (From Welches, W. R. and Baldwin, T. O., *Biochemistry*, 20, 512, 1981. With permission.)

used to "solubilize" insoluble proteins. Meighen and co-workers[21] have produced a "variant" form of bacterial luciferase through reaction with succinic anhydride. The succinylated protein retained the dimeric subunit structure of the native enzyme. By complementation experiments involving the mixing/hybridization of the modified and native enzyme, it was determined that succinylation of bacterial luciferase resulted in the inactivation of the α-subunit without markedly affecting the funcion of the β-subunit. Shetty and Rao[22] studied the reaction of succinic anhydride with arachin. In this study, reaction of the protein was performed in 0.1 M sodium phosphate, pH 7.8, with the pH maintained over the course of the reaction by the addition of 2.0 M NaOH. The extent of modification was determined by reaction of the unmodified primary amino groups on the protein with trinitrobenzene-sulfonic acid (see below). With a 200:1 molar excess of succinic anhydride, 82% of the available amino groups were succinylated with concomitant dissociation of the subunits of this protein. The reaction of chymotrypsinogen with succinic anhydride has been studied.[23] In these experiments, the reaction was performed under ambient conditions in 0.05 M sodium phosphate, pH 7.5. During the course of the reaction the pH was maintained at 7.5 by the addition of 1.0 M NaOH. Chymotrypsinogen (1 g) was dissolved in the sodium phosphate buffer and 50 mg of succinic anhydride was added over a 30-min period. Under these conditions, 8 of the 14 lysine residues were modified.

Citraconic anhydride has proved useful since the modification of lysine residues with this reagent is a reversible reaction. Reaction conditions for the modification of lysine residues in proteins are similar to those described above for other carboxylic acid anhydrides. Atassi and Habeeb[24] have discussed the use of this reagent in some detail. As an example, the reaction of egg white lysozyme with citraconic anhydride has been studied.[25] With multiple additions of reagent, all primary amino groups were modified at pH 8.2 (the pH of the reaction mixture was maintained with a pH-stat). The product of the reaction was heterogeneous as judged by polyacrylamide gel electrophoresis. All citraconyl groups could be removed by treatment with 1.0 M hydroxylamine at pH 10.0. This treatment also resulted in an electrophoretically homogeneous species. Complete removal of the citraconyl groups could also be achieved by incubation at pH 4.2 for 3 hr at 40°C.

More recently, reaction with citraconic anhydride has been used to dissociate nucleoprotein complexes.[26] Modification of the lysine residues with citraconic anhydride (pH 8.0 to 9.0 maintained with pH-stat) resulted in a marked change in the charge relationship between the ε-amino groups of lysine and the phosphate backbone of the nucleic acid, allowing subsequent separation of protein from nucleic acid (Figure 9). The citraconyl groups were subsequently removed from this protein by incubation at pH 3.0 to 4.0 at 30°C for 3 hr (Figure 10).

Mahley and co-workers have prepared the acetoacetyl derivatives of lipoproteins by reaction with diketene in 0.3 borate, pH 8.5.[27,28] The modification of tyrosyl and seryl residues also can occur under these conditions, but the O-acetoacetyl groups can be removed by dialysis against a mild alkaline buffer such as bicarbonate. The modification at lysyl residues can be reversed by 0.5 M hydroxylamine, pH 7.0 at 37°C. A 0.06 M solution of diketene was prepared by taking 50 μℓ diketene into 10 mℓ 0.1 M sodium borate, pH 8.5. The modification was performed at pH 8.5. The extent of modification was determined by subsequent titration with fluorodinitrobenzene. The effect of the modification of lysine residues on the in vivo clearance of lipoproteins in rats has been investigated.[28]

Urabe and co-workers[29] prepared various mixed carboxylic acid anhydrides of tetradecanoic acid and oxa derivatives which varied in their "hydrophobicity". This represented an attempt to change the surface properties of the enzyme molecule, in this case, thermolysin. The carboxylic acid anhydrides were formed *in situ* from the corresponding acid and ethyl-chloroformate in dioxane with triethylamine. The modification reaction was performed in 0.013 M barbital, 0.013 M CaCl$_2$, pH 8.5 containing 39% (v/v) dioxane and was terminated with neutral hydroxylamine which also served to remove O-acyl derivatives. The extent of reaction was determined by titration with trinitrobenzenesulfonic acid. Derivatives obtained with tetradecanoic acid and 4-oxatetradecanoic acid were insoluble. Derivatives obtained with 4,7,10-trioxatetradeconoic acid and 4,7,10,13-tetraoxotetradecanoic acid both had ap-

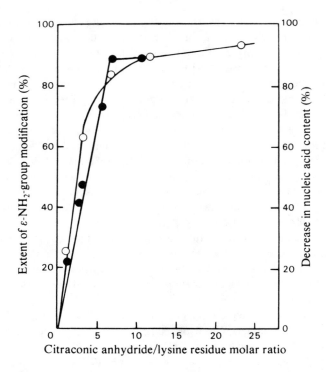

FIGURE 9.   The effect of citraconic anhydride concentration on
the reaction of citraconic anhydride with the ε-amino groups of lysine
residues (○) and the decrease in nucleic acid content, (●) in a
nucleoprotein complex prepared from yeast. The pH of the reaction
was maintained at pH 9.5 by the addition of 3.5 M NaOH. (From
Shetty, J. K. and Kinsella, J. E., *Biochem. J.*, 191, 269, 1980. With
permission.)

proximately seven amino groups modified per mole of enzyme, showed little if any loss in
either proteinase or esterase activity, and possessed enhanced thermal stability. Howlett and
Wardrop[30] were able to dissociate the components of human erythrocyte membrane by the
use of 3,4,5,6-tetrahydrophthalic anhydride. The reaction was performed in 0.02 M Tricine,
pH 8.5. The 3,4,5,6-tetrahydrophthalic anhydride was introduced into the reaction mixture
as a dioxane solution (a maximum of 0.10 mℓ/5 mℓ reaction mixture). The pH was maintained
at pH 8.0 to 9.0 with 1.0 M NaOH. The reaction was considered complete when no further
change in pH was observed. The extent of modification was determined by titration with
trinitrobenzenesulfonic acid. The reaction could be reversed by incubation for 24 to 48 hr
at ambient temperature following the addition of an equal volume 0.1 M potassium phosphate,
pH 5.4 (the final pH of the reaction mixture was 6.0).

The reaction of primary amino groups in proteins with cyanate (Figure 11) has been a
useful procedure for several decades. Stark and co-workers[31] pursued the observation that
ribonuclease was inactivated by urea in a time-dependent reaction. It was established that
this inactivation was a reflection of the content of cyanate in the urea preparation (Figure
12). This observation was subsequently developed into a method for the quantitative determination of amino-terminal residues in peptides and proteins.[32] The reaction of cyanate with
amino acid residues has been reported by Stark.[33] The ε-amino group of lysine is the least
reactive (k = 2.0 × 10^{-3} M^{-1} min^{-1}) as compared to the α-amino group of glycylglycine
(k = 1.4 × 10^{-1} M^{-1} min^{-1}). The carbamyl derivative of histidine is quite unstable as is
the corresponding derivative of cysteine. Concern should be given to reaction at residues

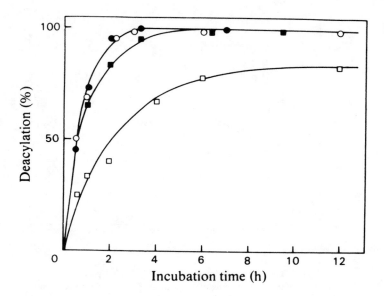

FIGURE 10. The effect of pH on the rate of deacylation of the ε-amino groups of lysine in the citraconylated protein prepared from the nucleoprotein complex prepared as described in Figure 9. The reactions were performed at 30°C at pH 3.0 (●), pH 4.0 (○), pH 5.0 (■) or pH 6.0 (□). (From Shetty, J. K. and Kinsella, J. E., *Biochem. J.*, 191, 269, 1980. With permission.)

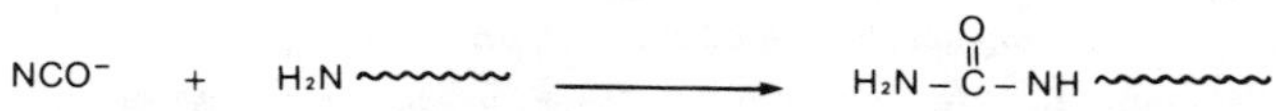

FIGURE 11. The reaction of primary amino groups with cyanate.

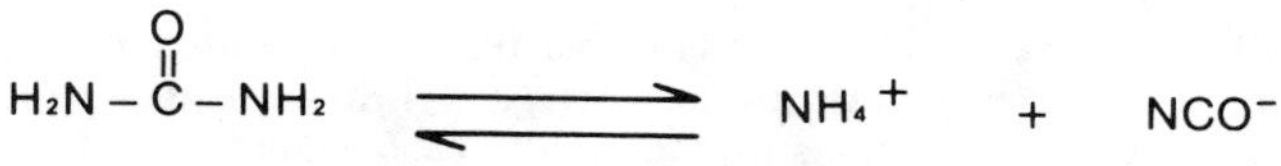

FIGURE 12. The formation of cyanate from urea.

other than amines. For example, the reaction of chymotrypsin with cyanate results in loss of catalytic activity associated with the carbamylation of the active site serine residue.[34]

More recently, Manning and co-workers[35-38] established that the modification of sickle cell hemoglobin with cyanate increased the oxygen affinity of this protein. As with the studies of Stark and co-workers described above, interest in the use of cyanate derived from consideration of the effect of urea.[35] The modification of primary amino groups in hemoglobin has been considered in some detail. It has been established that the amino-terminal value of hemoglobin is more reactive to cyanate in deoxygenated blood than in partially deoxygenated blood. At pH 7.4, the amino-terminal valyl residues of oxyhemoglobin S are carbamylated 50 to 100 times faster than lysyl residues.[36] The rate of incorporation is radiolabeled cyanate into oxyhemoglobin S is shown in Figure 13. Analysis of this reaction after 5 min showed 1 mol valine modified per tetramer and 0.27 mol homocitrulline (the reaction product of cyanate with the ε-amino group of lysine) per mole tetramer. After 30 min of reaction, 3 of the 4 amino-terminal valine residues are carbamylated and 2 of 44 (total) lysine residues are modified. The rates of reaction of various hemoglobins and separated chains with cyanate are shown in Figure 14. Figure 15 shows the separation of the α- and β-chain of carbamylated hemoglobin S. The same laboratory has examined the carbamylation of α-chain and β-chain

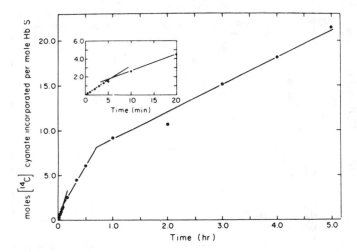

FIGURE 13.   The carbamylation of the α- and ε-amino groups of oxyhemoglobin S (sickle cell hemoglobin) as measured by the incorporation
of [¹⁴C] sodium cyanate at pH 7.4 in a pH-stat. Portions were removed at
the indicated periods of time, precipitated with cold 5% trichloroacetic
acid. The inset describes the early phase of the reaction demonstrating that
there are three distinct rates for the reaction. (From Lee, C. K. and Manning, J. M., *J. Biol. Chem.*, 248, 5861, 1973. With permission.)

*pH-dependent pseudo-first order rate constants of carbamylation of
hemoglobins with cyanate at pH 7.4 and 37°*

Carbonmonoxyhemoglobin solutions, 9.6 μM as tetramer, were
carbamylated with 40 mM NaN¹⁴CO. Deoxyhemoglobin solutions were prepared in the pH stat by mixing oxyhemoglobin
(final concentration 7.5 μM as tetramer) with about 1 mg of
Na₂S₂O₄ under a gentle stream of N₂ gas. After 5 min, NaN¹⁴CO
(final concentration 20 mM) was added for initiation of the reaction. The rate constants are an average of 4 determinations for
HbA and 6 determinations for HbS. The carbonmonoxy HMB-α
and HMB-β chains of Hb, 30 μM, were incubated with 20 mM
NaN¹⁴CO. The precision of the kinetic constants is ±0.50 for
the deoxyhemoglobins.

Protein	$k \times 10^2$ min⁻¹
CO-HbA	2.3
CO-HbS	2.4
Deoxy HbA	4.2
Deoxy HbS	4.0
CO-HMB-α_A	1.9
CO-HMB-α_S	1.6
CO-HMB-β_A	1.9
CO-HMB-β_S	1.6

FIGURE 14.   pH dependence of first-order rate constants for the carbamylation of hemoglobin preparations with cyanate at pH 7.4. Carbonmonoxyhemoglobin solutions (9.6 μM as tetramer) were carbamylated with radiolabeled sodium cyanate. Deoxyhemoglobin solutions were prepared as 7.5 μM
as tetramer with approximately 1 mg Na₂S₂O₄. (From Lee, C. K. and Manning,
J. M., *J. Biol. Chem.*, 248, 5861, 1973. With permission.)

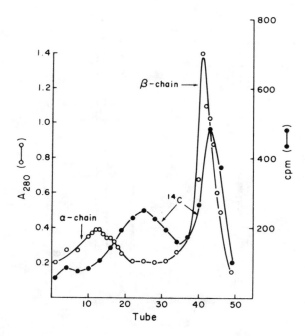

FIGURE 15. Countercurrent distribution patterns for carbamylated HbS (sickle cell hemoglobin). Oxygenated erythrocytes were incubated with 10 mM radiolabeled sodium cyanate for 1 hr at 37°C. Globin was prepared from the labeled erythrocytes and subjected to 50 transfers in 1% dichloroacetic acid-2-butanol. (From Lee, C. K. and Manning, J. M., *J. Biol. Chem.*, 248, 5861, 1973. With permission.)

in some detail.[37] With the deoxy protein, the ratio of radiolabel from ^{14}C-cyanate on α-chain as compared to the β-chain is 1.7:1.0 while it is 1:1 with the oxy protein. This analysis required the development of methodology to separate the various chains and derivative chains (Figure 16). As mentioned above, the carbamylation of the amino-terminal valine residues of hemoglobin is approximately 2.5-fold greater in partially deoxygenated media as compared to fully oxygenated media. Thus, it would appear the reactivity of the amino terminal valine is a sensitive index of conformational change.[38] It is also of interest that removal of Arg[141] (α) with carboxypeptidase B abolishes the enhancement of carbamylation observed with the removal of oxygen from hemoglobin.

Mahley and co-workers[27] used carbamylation to explore the role of lysyl residues in the binding of plasma lipoprotein to fibroblasts. The reaction was performed in 0.3 M sodium borate, pH 8.0. The extent of modification was determined in two ways. In the first, the modified protein was subjected to acid hydrolysis. The amount of homocitrulline, the product of the reaction of the ε-amino group of lysine with cyanate, was considered equivalent to the number of lysine residues modified. However homocitrulline is partially degraded on acid hydrolysis to produce lysine (17 to 30%). In order to obviate this difficulty, these investigators removed a portion of the modified protein and reacted it under denaturing conditions with 2,4-dinitrifluorobenzene, yielding an acid-stable derivative. The number of lysine residues modified was therefore the sum of free lysine and homocitrulline obtained on amino acid analysis following acid hydrolysis.

In an elegant study by Plapp and co-workers,[39] the modification of lysyl residues in bovine pancreatic deoxyribonuclease A by several different reagents, including cyanate, was examined as shown in Figure 17. The modification with cyanate is performed at 37°C in 1.0

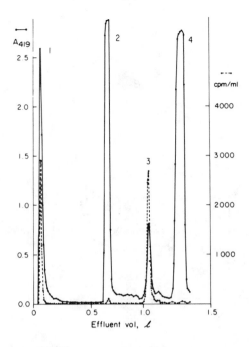

FIGURE 16.   The separation of ^{14}C carbamylated and noncarbamylated preparations of *p*-hydroxymercuribenzoate derivatives of the chains of hemoglobin S. The chromatographic medium used was carboxymethylcellulose (2.5 × 18 cm). Gradient elution was used with 850 mℓ of 10 m*M* potassium phosphate, pH 5.85 and a limit buffer of 850 mℓ of 15 m*M* K$_2$HPO$_4$, both solutions 1 m*M* with respect to EDTA. Peak 1 contained the carbamylated β-chain; peak 2, the noncarbamylated β-chain; peak 3, the carbamylated α-chain and peak 4, the noncarbamylated α-chain. (From Njikam, N., Jones, W. M., Nigen, A. M., Gillette, P. N., Williams, R. C., Jr., and Manning, J. M., *J. Biol. Chem.*, 248, 8052, 1973. With permission.)

M triethanolamine hydrochloride, pH 8.0. The extent of modification was determined by analysis for homocitrulline following acid hydrolysis. A time course of hydrolysis was utilized to provide for the accurate determination of homocitrulline since this amino acid slowly decomposes to form lysine during acid hydrolysis (see above). This modification was sensitive to the conformation of the protein since both the extent of modification and loss of catalytic activity depended on the presence or absence of calcium ions as shown in Figures 18 and 19.

Chollet and Anderson[40] have examined the modification of lysyl residues with potassium cyanate in the catalytic subunit of tobacco ribulosebiphosphate carboxylase. The modification was performed in 0.050 *M* HEPES, 0.025 *M* NaCl, pH 7.4. Stoichiometry was not established in this study but it was noted that modification occurred at both the amino terminal and the ε-amino groups of lysine.

The reaction of imidoesters with the primary amino groups of proteins has been the subject of considerable investigation in the past 10 to 20 years. The most extensive use of this class of reagents has been the covalent cross-linking of proteins (see Volume II, Chapter 5). These reagents have the particular advantage in that the charge of the lysine residue is maintained during the modification as shown for the reaction of lysine with methyl acetimidate in Figure 20. Ethyl acetimidate has been used to study the role of lysyl residues in thrombin.[41] The reaction was performed with a 1000-fold molar excess of reagent in 0.02 *M* sodium borate-

Derivative	Substituent	CaCl₂ used in preparation	No. of amino groups modified	Activity
		mM		%
Guanidino	$\oplus NH_2$ / $H_2N—C—$	0	8.8 ε, 0.1 α	40
		5	8.9 ε, 0.2 α	75
Picolinimidyl	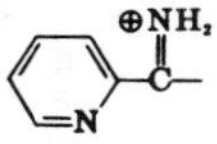 (pyridyl structure)	0	9.5 (ε + α)	100
		1	9 ε, 0.7 α	100
α-Picolinimidyl, ε-guanidino		5	9.0 ε, 0.9 α	65
Carbamyl	O / $H_2N—C—$	0	7 ε, 1 α	55ᵃ
		0 or 10	7 ε, 1 α	90
		0 or 10	9 ε, 1 α	0
Trinitrophenyl	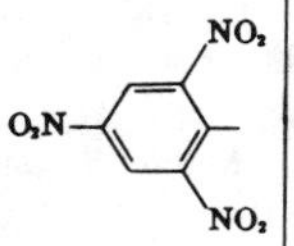 (trinitrophenyl structure)	0	1	95ᵃ
		0	7–8	0ᵃ
		5	4–5	100
		5	7	0

ᵃ Assayed in the absence of Ca⁺⁺.

FIGURE 17. The modification of bovine pancreatic DNase I by various reagents specific for the modification of lysine residues. The extent of lysine modification was determined by homocitrulline formation for reaction with *O*-methylisourea (guanidation), radiolabeled sodium cyanate for carbamylation and spectroscopy for picolinimydilation or trinitrophenylation. Enzymatic activity is expressed as a percent of that of a control preparation of DNAse. (From Plapp, B. V., Moore, S., and Stein, W. H., *J. Biol. Chem.*, 246, 939, 1971. With permission.)

0.15 M NaCl, pH 8.5. Amino acid analysis indicated that approximately 80% of the lysyl residues were modified under these conditions. The modification of a glutamine synthetase from *Bacillus stearothermophilus* with ethyl acetimidate has been studied by Sekiguchi and co-workers.[42] The modification was performed at pH 9.5 with 0.2 M phosphate for 1 hr at 35°C and terminated by dialysis at pH 7.2. The extent of modification was determined by titration of the modified protein with trinitrobenzenesulfonic acid. As these investigators suggest, consideration must be given to the possibility of cross-linking occurring with this reagent under the conditions used.[43] Monneron and d'Alayer[44] examined the reaction of either methyl acetamidate or dimethyl suberimidate with particulate adenylate cyclase. The reaction was performed in 0.05 M triethanolamine, 10% (w/v) sucrose, 0.005 M MgCl₂, pH 8.1. Plapp and co-workers[39] examined the reaction of methyl picolinimidate with pancreatic deoxyribonuclease. Methyl picolinimidate is an imidoester which reacts with the primary amino groups in proteins (Figure 21). The reaction was performed in 0.5 M triethanolamine hydrochloride, pH 8.0 containing 1 mM CaCl₂ with 0.1 M methyl picolinimidate for 22 hr at 25°C, then with 0.2 M methyl picolinimidate for an additional 8 hr. The extent of modification of a protein by methyl picolinimidate can be determined by spectral analysis (see Figure 22). Under these conditions, essentially all of the primary amino groups in deoxyribonuclease (nine lysine and one amino-terminal amino group) were modified but

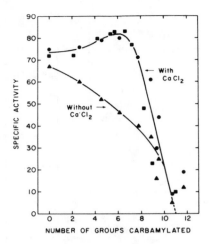

FIGURE 18.   The modification of bovine pancreatic DNase I with potassium cyanate in the presence (■) or absence of (●) 10 mM CaCl$_2$. The calcium-free DNase was assayed in the presence (●) or absence (▲) of 10 mM CaCl$_2$. (From Plapp, B. V., Moore, S., and Stein, W. H., *J. Biol. Chem.*, 246, 939, 1971. With permissin.)

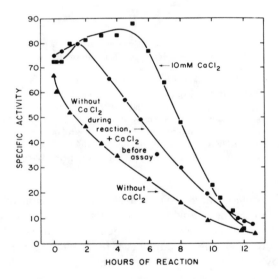

FIGURE 19.   The inactivation of bovine pancreatic DNase with cyanate. The reactions were performed at pH 8.0 (1.0 M triethanolamine) in the presence of 1.0 M potassium cyanate at 37°C. In the experiment described with the solid squares, 10 mM CaCl$_2$ was present during the reaction with potassium cyanate. In the experiment with the solid circles, the enzyme carbamylated in the absence of calcium ions was diluted into 10 mM CaCl$_2$ 15 min prior to assay as compared to the experiment described with the solid triangles where the DNase was carbamylated and assayed in the absence of CaCl$_2$. (From Plapp, B. V., Moore, S., and Stein, W. H., *J. Biol. Chem.*, 246, 939, 1971. With permission.)

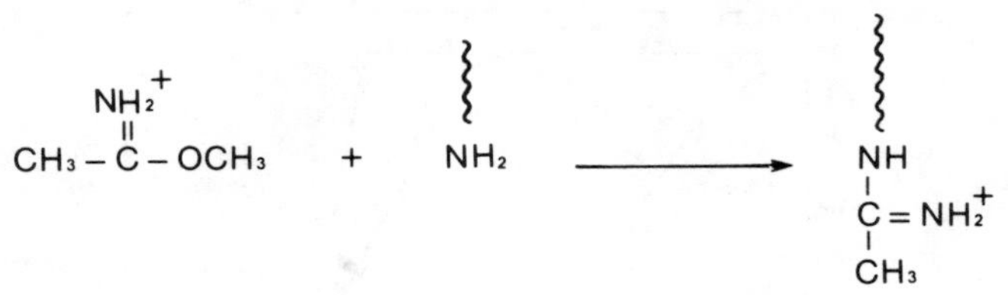

$$CH_3 - \overset{\overset{+}{N}H_2}{\underset{\|}{C}} - OCH_3 \quad + \quad NH_2 \quad \longrightarrow \quad \begin{matrix} NH \\ | \\ C = \overset{+}{N}H_2 \\ | \\ CH_3 \end{matrix}$$

FIGURE 20.   The reaction of lysine with methyl acetimidate.

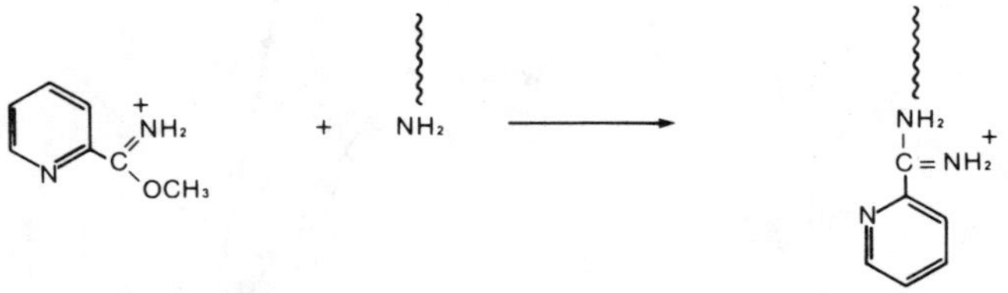

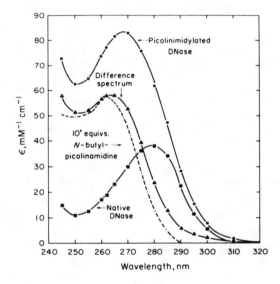

FIGURE 21.   The reaction of amino groups with methyl picolinimidate.

FIGURE 22.   The UV spectra of native DNase and the picolinimidylated derivative. The spectra of picolinimidylated DNase (●) and native DNase are presented (■) together with the difference spectrum (▲). From the absorbance at 262 nm and the extinction coefficient for *N*-butylpicolinamidine (5700 M^{-1} cm^{-1}), it was calculated that the modified enzyme contained 10 picolinimidyl groups (the theoretical difference spectrum for this extent of modification is shown by the dashed line.) The proteins were dissolved in 0.05 *M* sodium acetate, 1 m*M* $CaCl_2$ and clarified by centrifugation prior to analysis. A molecular weight of 31,000 was assumed in the calculations. (From Plapp, B. V., Moore, S., and Stein, W. H., *J. Biol. Chem.*, 246, 939, 1971. With permission.)

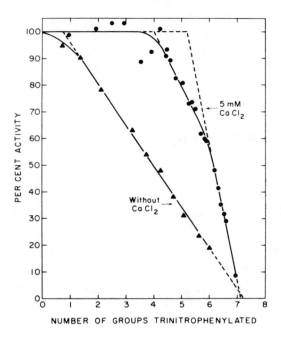

FIGURE 23.   The reaction of bovine pancreatic DNase by 2,4,6-trinitrobenzenesulfonic acid. Reaction was accomplished in 0.3 M sodium borate buffer, pH 9.5 in the dark for 5 hr in absence of added metal ion (▲) or for 23 hr in the presence of $5 mM$ CaCl$_2$(●). The extent of modification with 2,4,6-trinitrobenzenesulfonic acid was assessed by absorbance at 367 nm. (From Plapp, B. V., Moore, S., and Stein, W. H., *J. Biol. Chem.*, 246, 939, 1971. With permission.)

there was no change in biological activity. The work on DNase modification forms a particularly useful paper[39] because of its wealth of experimental detail as well as the comparison of the reaction of dexyribonuclease with four different reagents which modify primary amino groups: O-methyl isourea, methyl picolinimidate, cyanate, and trinitrobenzene sulfonic acid. As mentioned above, the reaction of methyl picolinimidate with deoxyribonuclease either in the presence or absence of calcium ions resulted in the modification of essentially all of the primary amino groups without change in biological activity. Reaction of deoxyribonuclease with cyanate in either the presence or absence of calcium ions eventually resulted in the modification of all of the primary amino groups with the complete loss of biological activity (see Figures 18 and 19). Modification of seven to eight amino groups with trinitrobenzenesulfonic acid resulted in the loss of all biological activity (Figure 23), while reaction of a similar number of residues with O-methylisourea only resulted in approximately 50% inactivation (see Figure 17). Plapp has also studied the reaction of methyl picolinimidate with horse liver alcohol dehydrogenase.[45] This study was somewhat unique in that modification of the enzyme resulted in enhanced catalytic activity reflecting more rapid dissociation of the enzyme-coenzyme complex. It should be noted that the derivatized lysine reverts back to lysine (60% yield) under the normal conditions of acid hydrolysis.

A number of investigators have used pyridoxal phosphate to modify lysyl residues in proteins. Pyridoxal phosphate is the cofactor form of vitamin B$_6$ and plays an important role in biological catalysis.[46] Pyridoxal phosphate is useful for the modification of lysine because of selectivity of reaction, spectral properties of the modified residue, reversibility of reaction,

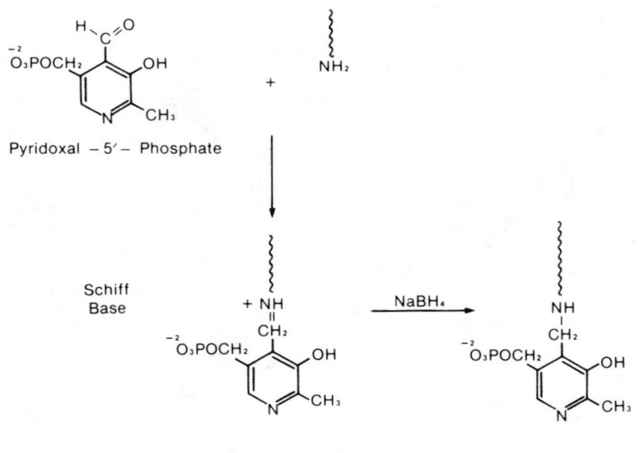

FIGURE 24.   The reaction of pyridoxal-5'-phosphate with amino groups in proteins.

Pyridoxal 5' – Phosphate   Pyridoxal   Pyridoxamine

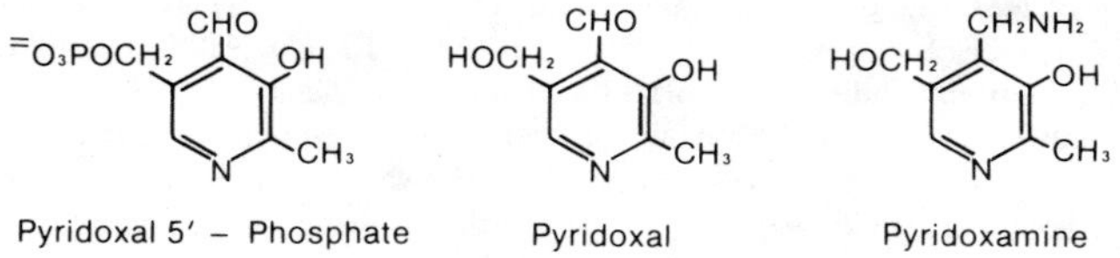

FIGURE 25.   The structure of pyridoxal-5'-phosphate, pyridoxal, and pyridoxamine.

and the establishment of sterochemistry by use of radiolabeled sodium borohydride (sodium borotritiide) to reduce the Schiff base initially formed on the reaction of pyridoxal phosphate with a primary amine. Pyridoxal phosphate will react with all primary amines (both ϵ-amino groups of lysine and the amino-terminal alpha amino function) in a protein (Figure 24). In general, pyridoxal 5'-phosphate is far more reactive than pyridoxal because of intramolecular hemiacetal formation and the neighboring group effect of the phosphate moeity (Figure 25). Horecker and co-workers investigated the reaction of pyridoxal phosphate with rabbit muscle aldolase.[47] The initial reaction produced a species with an absorbance maximum at 430 to 435 nm reflecting the protonated Schiff base form of the pyridoxal phosphate-protein complex. After reduction with sodium borohydride, the absorbance maximum is at 325 nm which is characteristic of the reduced Schiff base. This is a quite useful study in that the difference in reactivity between pyridoxal and pyridoxal-5'-phosphate is demonstrated as is the reversible nature of the initial complex. Schnackerz and Noltmann[48] compared the reaction of pyridoxal-5'-phosphate and other aldehydes in reaction with rabbit muscle phosphoglucose at pH 8.0. Pyridoxal-5'-phosphate (0.19 mM resulted in 82% inactivation while the following results were obtained with other aldehydes: pyridoxal (8.4 mM, 16% inactivation; acetaldehyde (75 mM), 75% inactivation; and acetone (75 mM), 31% inactivation. This last reaction is of interest as many investigators are unaware that acetone can react with amino groups in proteins. The reaction of acetone with primary amino groups has been known for some time[49] and is discussed in further detail below within the topic of reductive alkylation. The reaction of ribulose 1,5-bisphosphate carboxylase/oxygenase with pyridoxal-5'-phosphate has been studied by Paech and Tolbert.[50] Pyridoxal-5'-phosphate inactivated the enzyme with or without reduction with NaBH$_4$. This reaction was performed in 0.1 M Bicine (N,N-(2-hydroxyethyl) glycine), 0.010 M MgCl$_2$, 0.2 mM EDTA, 0.001 M diethiothreitol (see

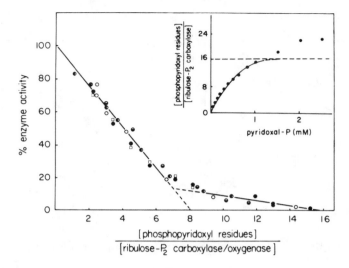

FIGURE 26.   The reaction of ribulose-1,5-bisphosphate carboxylase/
oxygenase with pyridoxal-5'-phosphate. The enzyme was incubated in 0.1
M N,N-bis(2-hydroxyethyl) glycine (BICINE), 10 m*M* NaCl$_2$, 0.2 m*M*
EDTA, 1 m*M* dithiothreitol, pH 8.0 with increasing quantities of pyridoxal-
5'-phosphate followed by reduction with sodium borohydride. Ribulose-
1,5-bisphosphate oxygenase activity (□) is shown for those proteins having
carboxylase activity (●). The inset shows the number of phosphopyridoxal
residues incorporated per enzyme moleule as function of pyridoxal-5'-
phosphate concentration. The extinction coefficient for the phosphopyridoxal derivative at 325 nm (the sodium borohydride reduced derivative)
was calculated to be 4,800 *M*$^{-1}$ cm^{-1}. (From Paech, C. and Tolbert, N.
F., *J. Biol. Chem.*, 253, 7864, 1978. With permission.)

Figure 26). The reaction demonstrated a optimum at pH 8.4. Spectral studies showed the
formation of a species absorbing at 432 n*M*. As is characteristic for the Schiff base derivative,
this peak disappears on reduction to yield a species with a optimum at 325 nm ($\Delta\epsilon$ = 4800
M^{-1} cm^{-1}). This supports the suggestion that the loss of activity observed on reaction with
pyridoxal-5'-phosphate is due to the formation of a Schiff base which can be reduced with
NaBH$_4$ to form a stable derivative, as opposed to the formation of a 2-azolidine ring with
a second nucleophile as has been observed by other investigators.[51-53] Jones and Priest[54]
have investigated the modification of apo-serine hydroxymethyltransferase with pyridoxal
phosphate and the subsequent use of the enzyme-bound pyridoxal phosphate as a structural
probe. Cortijo and co-workers[55] have suggested the use of the ratio of absorbance at 415
nm and 335 nm of enzyme-bound pyridoxal phosphate as an indication of the polarity of
the medium. Cake and co-workers[56] have demonstrated that modification of activated hepatic
glucorticoid receptor with pyridoxal-5'-phosphate obviated the binding of the receptor to
DNA. Greatly reduced inhibition was seen with pyridoxamine-5'-phosphate, pyridoxamine,
or pyridoxine (see Figure 27). Inhibition could be reversed by gel filtration or treatment
with dithiothreitol while treatment with NaBH$_4$ resulted in irreversible inhibition of DNA
binding. These investigators used 0.2 *M* borate, 0.25 *M* sucrose, 0.003 *M* MgCl$_2$ (pH 8.0)
as the solvent for reaction with pyridoxal-5'-phosphate. Slebe and Martinez-Carrion[57] have
introduced the use of phosphopyridoxal trifluoroethyl amine as a probe for pyridoxal phosphate binding sites in enzymes (Figure 28). Nishigori and Toft[58] explored the reaction of
pyridoxal-5'-phosphate with the avian progesterone receptor. Reaction with pyridoxal-5'-
phosphate was performed in 0.02 *M* barbital, 10% (v/v) glycerol, 0.005 m*M* dithiothreitol,
0.010 *M* KCl, pH 8.0. The modification was stabilized by NaBH$_4$. It is of interest that these

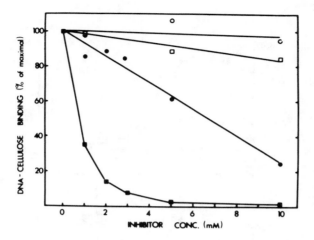

FIGURE 27. The specificity of the effect of pyridoxal-5'-phosphate on the DNA binding site of activated hepatic glucocorticoid receptor. The reactions were performed in 0.2 M boric acid, 0.25 M sucrose, 3 mM NaCl$_2$, p l8.0 at 0°C. The reactions included either pyridoxal-5'-phosphate, 0.75 mM (■), 6.5 mM pyridoxal (●), pyridoxamine-5'-phosphate (□), or pyridoxamine (○). In data not shown, pyridoxine or phosphate ions were without effect in the inhibition of DNA binding by the activated receptor. (From Cake, M. A., DiSorbo, D. M., and Litwack, G., *J. Biol. Chem.*, 253, 4886, 1978. With permission.)

investigators noted that the modification was readily reversed in Tris buffer unless stabilized by NaBH$_4$. Sugiyama and Mukohata[59] observed that modification with pyridoxal-5'-phosphate of the lysine residue in chloroplast coupling factor using 0.020 M Tricine, 0.001 M EDTA, 0.010 M MgCl$_2$, pH 8.0 resulted in complete inactivation of the ATPase activity. Peters and co-workers[60] reported on the inactivation of the ATPase activity in a bacterial coupling factor by reaction with pyridoxal-5'-phosphate. The modification was performed in 0.050 M morpholinosulfonic acid, pH 7.5. The inhibition was readily reversed by dilution or by 0.01 M lysine and was, as expected, stabilized by NaBH$_4$. Gould and Engel[61] reported on the reaction of mouse testicular lactate dehydrogenase with pyridoxal-5'-phosphate in 0.050 M sodium pyrophosphate, pH 8.7 at 25°C. This reaction resulted in the inactivation of the dehydrogenase activity. The inactivation was reversed by cysteine (Figure 29) and stabilized by NaBH$_4$. These investigators reported that the observed absorption coefficient at 325 nm may be decreased much as 50% with protein-bound pyridoxal phosphate. Thus, estimation of the number of lysine residues modified using the absorption coefficient obtained with model compounds might provide only a minimum value. Ogawa and Fujioka[62] studied the reaction of pyridoxal-5'-phosphate with saccharopine dehydrogenase in 0.1 M potassium phosphate, pH 6.8, at ambient temperature in the dark. Both spectral analysis (Figure 30) and tritium incorporation from sodium borotritiide reduction (Figure 31) were consistent with the modification of one lysine residue per mole of enzyme being responsible for the loss of enzyme activity. A value of $1 \times 10^4 \ M^{-1} \ cm^{-1}$ for the extinction coeffcient at 325 nm[63] was used in this study. It is of interest that this study demonstrated that it is possible to establish an equilibrium between the native and modified forms of the enzyme. The reversibility of the modification is shown in Figure 32. Also shown in Figure 32 is a series of experiments designed to determine the equilibrium constant for the reaction using a graphical method where the reciprocal of the concentration of pyridoxal-5-phosphate is plotted vs. the activity at equilibrium (Aeq) divided by the value obtained (Ao-Aeq) by

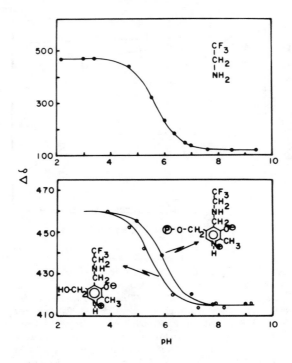

FIGURE 28.   The pH dependence of the chemical shift response of ^{19}F nuclear magnetic resonance with fluorinated compounds in the absence of enzymes. The top panel shows 2,2,2-trifluoroethylamine (100 mM). The solid curve indicates a theoretical titration curve of a simple ionization group for a pKa of 5.65. The bottom panel shows phosphopyridoxal trifluoroethyl amine (10 mM) before (●) and after (○) treatment with alkaline phosphatase. The solid lines show theoretical titration curves for a single ionization with a pK_a of 5.90 and 5.50 respectively. (From Slebe, J. C. and Martinez-Carrion, M., *J. Biol. Chem.*, 253, 2093, 1978. With permission.)

subtracting the activity at equilibrium (Aeq) from the initial activity (Ao). The slope of this graph provides the equilibrium constant (Keq) for the reaction under the experimental conditions (0.1 M potassium phosphate, pH 6.8 at 0°C). A value of $3.3 \times 10^3\,M^{-1}$ was obtained for the equilibrium constant as reaction is also shown in Figure 33. Protection is not provided by α-ketoglutarate in the absence of the reduced coenzyme. Pyridoxal was much less effective than pyridoxal-5'-phosphate in the inactivation of saccharopine dehydrogenase. The concentration of pyridoxal and pyridoxal-5'-phosphate were determined spectrophometrically in 0.1 M NaOH using an extinction coefficient of $5.8 \times 10^3\,M^{-1}\,cm^{-1}$ at 300 nm and $6.6 \times 10^3\,M^{-1}\,cm^{-1}$ at 388 nm respectively.[64] As described above, amine compounds have the potential to interfere in the reaction of pyridoxal-5'-phosphate with proteins. Moldoon and Cidlowski[65] demonstrate that 0.1 M Tris, pH 7.4 markedly interfered with the modification of rat uterine estrogen receptor with pyridoxal-5'-phosphate. These investigators also noted that, as in the other studies, 0.05 M lysine would block the modification reaction and could also reverse the modification if the Schiff base had not been reduced. Stock solutions of pyridoxal phosphate were prepared in 0.01 M NaOH to avoid acid decomposition. The importance of local environmental factors in the specificity of modification by pyridoxal phosphate is emphasized by Ohsawa and Gualerzi.[66] These investigators examined the modification of *Escherichia coli* initiation factor by pyridoxal phosphate in 0.020 M triethanol-

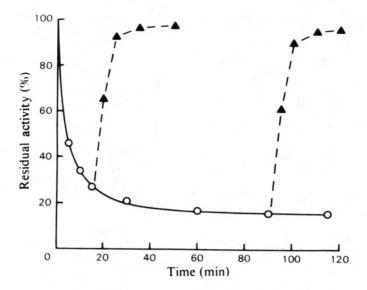

FIGURE 29.   The L-cysteine reversal of the inactivation of mouse
C_4 lactate dehydrogenase inactivated by pyridoxal-5′-phosphate.
Mouse C_4 lactate dehydrogenase was incubated with 1 m*M* pyri-
doxal-5′-phosphate in the dark at 25°C in 0.05 *M* sodium pyrophosphate, pH 8.7 (○). At 15 min and 90 min, 0.2 m*ℓ* portions
of the reaction mixture were removed and mixed with 10 μ*ℓ* of
1.0 *M* cysteine and assayed for enzyme activity at the time points
indicated (▲). (From Gould, K. G. and Engel, P. C., *Biochem.
J.*, 191, 365, 1980. With permission.)

amine, 0.03 *M* KCl, pH 7.8. In the course of the studies, it was observed that pyridoxal
phosphate will not react with poly (AUG). These investigators also reported the preparation
of N^6-pyridoxal lysine by reaction of pyridoxal phosphate with polylysine in 0.01 *M* sodium
phosphate, pH 7.2 at 37°C followed reduction with $NaBH_4$. The reduction was terminated
by the addition of acetic acid. Acid hydrolysis (6 *N* HCl, 110°C, 22 hr) yielded N^6-pyridoxal-
L-lysine. Bürger and Görisch[67] reported the inactivation of histidinol dehydrogenase upon
reaction with pyridoxal phosphate in 0.02 *M* Tris, pH 7.6. This modification could be
reversed by dialysis unless the putative Schiff base was stabilized by reduction with NaH_4
(*n*-octyl alcohol added to prevent foaming). These investigators used a $\Delta\epsilon$ for ϵ-amino
pyridoxal lysine of $1 \times 10^4 \ M^{-1} \ cm^{-1}$ at 325 nm.

   The modification of primary amines in proteins by reductive alkylation has proved to be
a useful reaction (Figure 34). This reaction has the advantage that the basic charge properties
of the modified residue are preserved. The early work on this modification has been reviewed
by Feeney, Means, and co-workers.[68-70] Both monosubstituted and disubstituted derivatives
can be prepared depending upon reaction conditions and the nature of the carbonyl compound.[69]

   Rice and co-workers[71] reported the stabilization of trypsin by reductive methylation. This
reaction utilized formaldehyde/sodium borohydride in 0.2 *M* sodium borate, pH 9.2 in the
cold. Unsubstituted amino groups were present after the reaction as demonstrated by titration
with trinitrobenzenesulfonic acid. The amino-terminal isoleucine residue was not modified
under these conditions. Morris and co-workers[72] investigated the reductive methylation of
monellin. The modification was performed in 0.2 *M* sodium borate, pH 8.0 with 1 m*M*

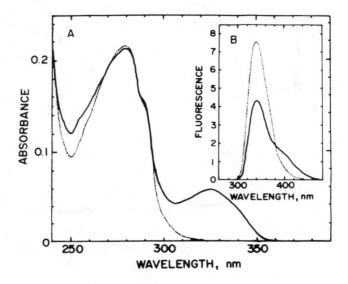

FIGURE 30.   The UV spectra and fluorescence spectra of native and pyridoxal-5'-treated saccharopine dehydrogenase (L-lysine forming). The enzyme preparation (17.9 nmol) was incubated with 0.5 mM pyridoxal-5'-phosphate in 0.1 M potassium phosphate, pH 6.8 followed by reduction with sodium borohydride and dialysis again with 0.1 M potassium phosphate, pH 6.8 in the dark. Panel A shows the UV absorption spectra for the native (dotted line) and modified enzyme (solid line). Panel B shows the fluorescence emission spectra for the native (dotted line) and the reduced enzyme (solid line). The excitation wavelength in the fluorescence spectra was 280 nm. The protein concentrations for the native and modified enzyme were identical in these experiments. (From Ogawa, H. and Fujioka, M., *J. Biol. Chem.*, 255, 7420, 1980. With permission.)

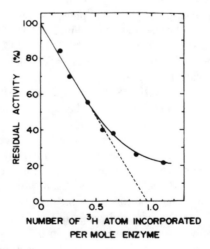

FIGURE 31.   Correlation between the loss of saccharopine dehydrogenase enzyme activity and the extent of modification with pyridoxal-5'-phosphate after reduction with tritiated sodium borohydride. (From Ogawa, H. and Fujioka, M., *J. Biol. Chem.*, 255, 7420, 1980. With permission.)

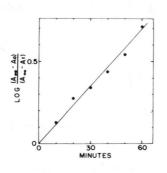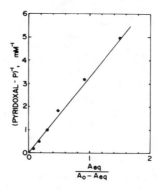

FIGURE 32.   Left panel shows recovery of activity of saccharopine
dehydrogenase inactivated by pyridoxal-5′-phosphate. The enzyme (23
nmol) was inactivated by reaction in 0.2 mℓ of 0.1 M potassium
phosphate buffer, pH 6.8 at 0°C. After 30 min, 10 μℓ portions were
transferred to 1.7 mℓ 0.1 potassium phosphate, pH 6.8 at 0°C. At the
indicated times, portions were reduced with sodium borohydride and
assayed for enzyme activity. A_o indicates enzyme activity at zero time,
A_t indicates enzyme activity at time t while A_{eq} indicates enzyme
activity at equilibrium. The right panel shows a plot of the reciprocal
of pyridoxal-5′-phosphate concentration versus $A_{eq}/A_o - A_{eq}$. Saccar-
opine dehydrogenase (2 nmol) was incubated in 0.1 M potassium phosphate, pH 6.8 and residual enzyme activity determined at 24°C. (From
Ogawa, H. and Fujioka, M., *J. Biol. Chem.*, 255, 7420, 1980. With
permission.)

monellin (11 mM) with respect to primary amino groups in the cold. Sodium borohydride
was added to give a final concentration of 0.5 mg/mℓ, and 1 to 5 μℓ of 6 to 8 M formaldehyde
was added per mℓ of solution. Tritiated formaldehyde was used to establish the extent of
modification. One of the problems with the use of formaldehyde in this reaction is the
presence of paraformaldehyde. Chen and Benoiton[73] obviated this difficulty by the *in situ*
generation of formaldehyde from methanol.

The introduction of sodium cyanoborohydride as a reducing agent for this reaction represented a real advance. Sodium cyanoborohydride is stable in aqueous solution at pH 7.0.
Unlike sodium borohydride which can reduce aldehydes and disulfide bonds, sodium cyanoborohydride only reduces the Schiff base formed in the initial process of reductive
alkylation. The radiolabeling of proteins using ^{14}C-formaldehyde and sodium cyanoborohydride has been reported.[74] The modification was performed in 0.04 M phosphate, pH 7.0
at 25°C. The modification can be performed equally well at 0°C but, as would be expected,
takes a longer period of time; there is no effect on the extent of the modification (see Figure
35). In this regard, these authors estimated that the same extent of modification obtained in
1 hr at 37°C could be achieved in 4 to 6 hr at 25°C or 24 hr at 0°C. Although the majority
of experiments in this study were performed in phosphate buffer at pH 7.0, equivalent results
can be obtained in Tris or HEPES (N-2-hydroxyethylpiperazine-N'-2-ethanesulfonic acid)
buffer at pH 7.0. A greater extent of modification was observed with sodium cyanoborohydride at pH 7.0 than with sodium borohydride at pH 9.0. The effect of carbonyl compounds
of different size on the extent of reductive alkylation has been examined by Feeney and co-
workers.[75] Of particular value in this study was the reporting of the elution positions of the
various derivatives of lysine during amino acid analysis (Figure 36). Thus, ε-dimethyl lysine
and ε-monomethyl lysine elute prior to lysine, while ε-isopropyl lysine elutes between lysine
and histidine, ε-dibutyl lysine between histidine and ammonia, ε-butyl lysine and ε-cyclo-
pentyl lysine between ammonia and arginine, while both the ε-cyclohexyl and ε-benzyl

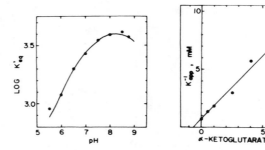

FIGURE 33.   The left panel shows the effect of pH on the equilibrium constant for the inactivation of saccharopine dehydrogenase by pyridoxal-5'-phosphate. The enzyme (2.5 nmol) was incubated at 24°C with 0.1 M potassium phosphate at the indicated pH. Values for the equilibrium constant were calculated from the equation

$$K_{eq} = (A_o - A_{eq})/A_{eq}[P]$$

The points were determined experimentally and the line was calculated from the following equation:

$$\log K'_{eq} = \log K + \log\left(1 + \frac{[H^+]}{K_{3S}} + \frac{[H^+]^2}{K_{2S}\,K_{3S}} + \frac{[H^+]^3}{K_{1S}\,K_{2S}\,K_{3S}}\right)$$
$$- \log\left(1 + \frac{[H^+]}{K_{3P}} + \frac{[H^+]^2}{K_{2P}\,K_{3P}} + \frac{[H^+]^3}{K_{1P}\,K_{2P}\,K_{3P}}\right) - \log\left(1 + \frac{[H^+]}{K_E}\right)$$

The right panel shows a plot of reciprocal apparent equilibrium constant vs. α-ketoglutarate concentration. The enzyme (2 nmol) was incubated with 1.0 mM pyridoxal-5'-phosphate containing 0.2 mM reduced nicotinamide-adenine dinucleotide (NADH) and α-ketogluarate at the concentrations indicated. The values for the equilibrium constants were calculated using the equation described above. (From Ogawa, H. and Fujioka, M., *J. Biol. Chem.*, 255, 7420, 1980. With permission.)

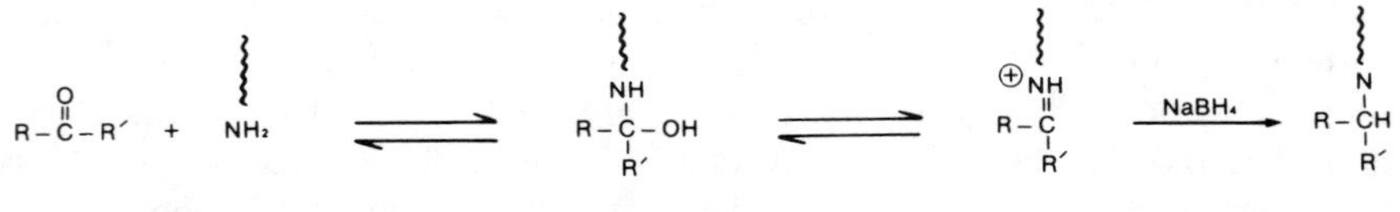

FIGURE 34.   The reductive alkylation of amino groups in proteins.

derivatives of lysine elute after arginine. The extent of modification is more a reflection of the type of alkylating agent and reaction conditions than an intrinsic property of the protein under study. For example, nearly 100% disubstitution can be obtained with formaldehyde and approximately 35% disubstitution with n-butanol, while only monosubstitution can be obtained with acetone, cyclopentanone, cyclohexanone, and benzaldehyde. While most of the products of reductive alkylation retained solubility, the reaction products obtained with cyclohexanone and benzaldehyde tended to precipitate.

In another study, the reversible reductive alkylation of proteins has been examined.[76] Both glycolaldehyde and acetol will react with the primary amino groups in proteins to yield derivatives which can be cleaved with periodate under mild basic condition to yield the free amine. Figure 37 shows the distribution of reaction products of lysine and glycolaldehyde as a function of pH with either sodium borohydride (A) or sodium cyanoborohydride (B) as the reducing agent. It is apparent that sodium cyanoborohydride is much more effective

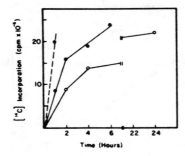

FIGURE 35. The reductive alkylation of chymotrypsinogen A. The reaction was performed in 0.04 *M* potassium phosphate buffer. Five μℓ of [14C] formaldehyde was added to 300 μℓ of a 1 mg/mℓ solution of chymotrypsinogen A followed by the addition of freshly prepared sodium cyanoborohydride (6 mg/mℓ in 0.04 *M* potassium phosphate, pH 7.0). The reactions were performed at 0°C (○), 25°C(◑) or 35°C(●). Portions were removed at the indicated times and radiolabel incorporation was determined. (From Dottavio-Martin, D. and Ravel, J. M., *Analyt. Biochem.*, 87, 562, 1978. With permission.)

MODIFICATION OF AMINO GROUPS

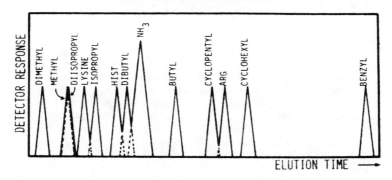

FIGURE 36. The relative elution positions for various lysine derivatives obtained by reductive alkylation. The dimethyl and methyl derivatives were obtained by using formaldehyde. The isopropyl derivatives were obtained by reaction with acetone. The butyl derivatives were obtained by reaction with butanol. The cyclopentyl derivative was obtained by reaction with cyclopentanone. The cyclohexyl derivative was obtained by reaction with cyclohexanone. The benzyl derivative was obtained by reaction with benzyldehyde. The reductive alkylation was performed in the presence of sodium borohydride with the inclusion of *p*-dioxane to facilitate the solution of the carbonyl reagents. The elution positions of lysine, histidine(HIS), and arginine(ARG) are included for reference. The analysis was performed on a Technicon Autoanalyzer using a sigmoidal gradient with an initial solvent of 0.05 *M* sodium citrate, pH 3.8, to a limit buffer of 0.05 *M* sodium citrate, 0.6 *M* sodium chloride, pH 5.1 which contained 30 mℓ isopropyl alcohol (for the butyllysine derivative), 49 mℓ isopropyl alcohol (for the isopropyllysine derivative), or 85 mℓ isopropyl alcohol (for the methyl-, cyclopentyl-, cyclohexyl-, and benzyllysine derivatives). (From Fretheim, K., Iwai, S., and Feeney, R. F., *Int. J. Peptide Protein Res.*, 14, 451, 1979. With permission.)

in the range of pH 6.0 to pH 8.0 while sodium borohydride is more effective under more alkaline conditions. Treatment of 30.0 mg lysozyme in 6.0 mℓ 0.2 *M* sodium borate, pH 9.0, with 60 mg glycoladehyde and 10 mg sodium borohydride at ambient temperature resulted in 60% 2-hydroxyethylation. Treatment of 20 mg ovomucoid in 2.0 mℓ 0.2 *M*

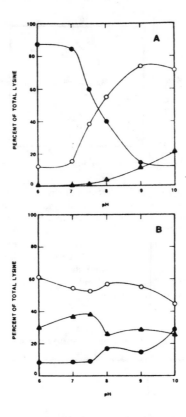

FIGURE 37.   The pH dependence of the reductive alkylation of α-L-lysine using two different reducing agents. In panel A sodium borohydride was used as the reducing agent while in panel B sodium cyanoborohydride was used as the reducing agent. The composition of the reaction products was determined after removal of the α-*N*-acetyl group by acid hydrolysis. The products shown are lysine (●), ε-*N*-(2-hydroxyethyl)lysine (○), and ε-*N,N*-bis-(2-hydroxyethyl)lysine (▲). (From Geoghegan, K. F., Cabacungan, J. C., Dixon, H. B. F., and Feeney, R. E., *Int. J. Peptide Protein Res.*, 17, 345, 1981. With permission.)

sodium borate, pH 9.0 with 10% acetol and 30 mg sodium borohydride (added in portions) resulted in 55% hydroxyisopropylation. In both situations, the reaction was terminated by adjustment of the pH to 5 with glacial acetic acid. The extent of modification was determined either by titration with trinitrobenzenesulfonic acid and/or by amino acid analysis after acid hydrolysis. Periodate oxidation could be accomplished with 0.015 *M* sodium periodate at pH 7.9 for 30 min at ambient temperature. Figure 38 describes the treatment of 2-hydroxyethylated lysozyme with periodate. Lysozyme with 75% of the primary amino groups modified was allowed to react at ambient temperature with 0.018 *M* sodium periodate in 0.2 *M* sodium phosphate, pH 7.0. The loss of activity occurring during the reaction is a reflection of the oxidation of the catalytically essential tryptophanyl residue in this protein.

As mentioned above, the replacement of sodium borohydride with sodium cyanoborohydride appears to represent a significant advance in the stabilization of Schiff bases in proteins after reaction with carbonyl compounds. There are several difficulties associated with the use of sodium borohydride in this reaction including the reduction of aldehydes to alcohols, the dependence of reduction on pH, the reduction of disulfide bonds, and the possible cleavage of peptide bonds. Jentoft and Dearborn have studied the use of sodium

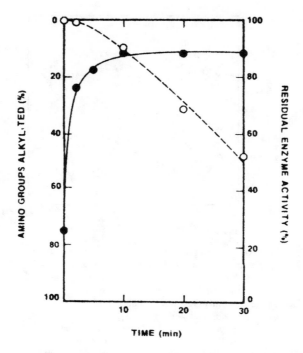

FIGURE 38.   The reaction of lysozyme modified by reductive alkylation with glycolaldehyde, with sodium periodate. The 2-hydroxyethylated protein (2 mg/mℓ) was treated with 15 mM sodium periodate in 0.1 M sodium phosphate, pH 7.0. The reaction was monitored by % amino groups remaining (●) by amino acid analysis and catalytic activity (○). (From Geoghegan, K. F., Cabacungan, J. C., Dixon, H. B. F., and Feeney, R. E., *Int. J. Peptide Protein Res.*, 17, 345, 1981. With permission.)

cyanoborohydride in some detail.[77] In particular, the preparation of sodium cyanoborohydride is critical and most, if not all, commercial preparations require recrystallization prior to use. This reflects the presence of impurities which limit the extent of the reductive alkylation (see below). Recrystallization is accomplished by dissolving 11 g of sodium cyanoborohydride in 25 mℓ acetonitrile. Insoluble material is removed by centrifugation. Crystallization is accomplished by the addition of 150 mℓ methylene chloride and allowing to stand overnight at 4°C. The recrystallized sodium cyanoborohydride is collected by filtration and stored in a vacuum desiccator. A fresh solution of reagent is prepared daily. Using [14C]-formaldehyde and sodium cyanoborohydride, the major product is ε-methylated lysine, with minor incorporation of radiolabel into arginine and histidine. Figure 39 shows the effect of sodium cyanoborohydride incorporation on the extent of reductive methylation (14C-formaldehyde) in 0.1 M HEPES buffer, pH 7.5. The rate of reductive methylation in this experiment with albumin was not sensitive to cyanoborohydride concentration. The effect of pH on the reaction is shown in Figure 40. Optimal reductive methylation was obtained at pH values greater than 8.0 during a short-term (10 min) incubation. The effect of pH is much less pronounced at longer periods of incubation (1 to 2 hr) with optimal reductive methylation occurring between pH 7.0 and pH 8.0. The effect of temperature on the rate of reaction at pH 7.5 (0.1 M HEPES) is shown in Figure 41. Note that although the reaction is much slower at 4°C, the extent of modification is almost equivalent to that achieved at higher temperature. The effect of formaldehyde concentration and protein concentration on the extent of reductive

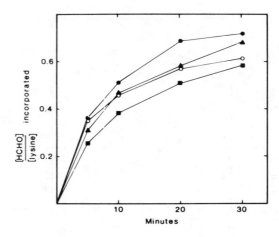

FIGURE 39. The effect of sodium cyanoborohydride concentration on the rate of reductive methylation of albumin. The reaction mixtures contained 1 mg/mℓ of albumin, 2 nm[¹⁴C] formaldehyde in 0.1 *M* HEPES buffer, pH 7.5. The reaction mixtures were maintained at 22°C and terminated at the indicated points by the addition of trichloroacetic acid and incorporated radiolabel determined. The concentration of sodium cyanoborohydride used in the experiment was 10 m*M* (■), 25 m*M* (▲), 100 m*M* (●) or 250 m*M* (○). (From Jentoft, N. and Dearborn, D. G., *J. Biol. Chem.*, 254, 4359, 1979. With permission.)

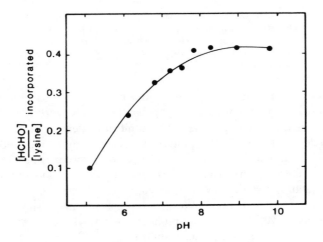

FIGURE 40. The pH dependence for the reductive methylation of lysine residues in albumin. The reaction mixtures contained 0.86 mg/mℓ of protein, 20 m*M* sodium cyanoborohydride, and 2 m*M* [¹⁴C] formaldehyde and were incubated for 10 min at 22°C. The extent of modification was determined by incorporation of radiolabel. The buffers used were sodium acetate (pH 5.1), sodium phosphate (pH 6.1 and pH 6.8), HEPES (pH 7.2 and 8.25), and sodium borate (pH 9.0 and 9.8). (From Jentoft, N. and Dearborn, D. G., *J. Biol. Chem.*, 254, 4359, 1979. With permission.)

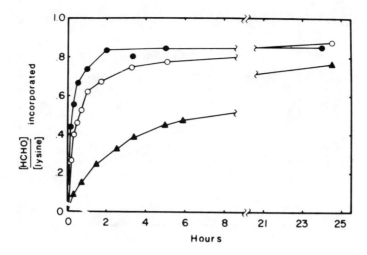

FIGURE 41. The effect of time and temperature on the reductive methylation of albumin. The reaction mixture contained 1.0 mg/mℓ albumin in 0.2 *M* HEPES, pH 7.5 in the presence of 20 m*M* sodium cyanoborohydride, and 2 m*M* [^{14}C] formaldehyde. The reactions were incubated at 4°C (▲), 22°C (○), or 37°C (●) for the indicated periods of time. (From Jentoft, N. and Dearborn, D. G., *J. Biol. Chem.*, 254, 4359, 1979. With permission.)

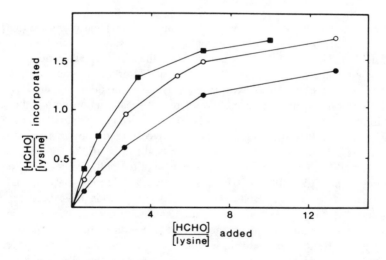

FIGURE 42. The effect of formaldehyde concentration and protein concentration on the reductive methylation of albumin. The reaction mixtures contained 20 m*M* sodium cyanoborohydride, with varying amounts of formaldehyde and albumin in 0.1 *M* HEPES buffer, pH 7.5. In (●), the reaction mixture contained 0.43 mg/mℓ albumin, in (○), 1.07 mg/mℓ albumin and in (■), 3.41 mg/mℓ albumin. The reactions were maintained at 22°C for 2 hr. The extent of modification was determined by measuring the extent of radiolabel incorporated [^{14}C] formaldehyde. (From Jentoft, N. and Dearborn, D. G., *J. Biol. Chem.*, 254, 4359, 1979. With permission.)

methylation is shown in Figure 42. An examination of this data suggests that approximately 80% modification can be achieved at a protein (bovine serum albumin) concentration of greater than 1 mg/mℓ at a formaldehyde/lysine ratio of 8 while virtually quantitative modification was obtained at a ratio of 12. These investigators also noted that Tris, β-mercap-

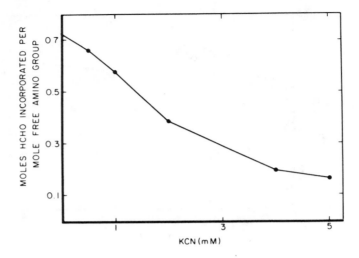

FIGURE 43.   The effect of the addition of potassium cyanide (KCN) on the reductive methylation of albumin. Reaction mixtures contained 1.04 mg/mℓ bovine serum albumin, 2 mM [^{14}C] formaldehyde, 10 mM sodium cyanoborohydride in 0.05 M HEPES, pH 7.5 with varying amounts of KCN as indicated. The extent of modification was determined by measuring the amount of radiolabel incorporation. (From Jentoft, N. and Dearborn, D. G., *Analyt. Biochem.*, 106, 186, 1980. With permission.)

toethanol, dithiothreitol, ammonium ions (as ammonium sulfate) and guanidine (5 M) inhibited the reductive alkylation of albumin by formaldehyde and sodium cyanoborohydride in 0.1 M HEPES, pH 7.5. The use of [^{13}C] formaldehyde in the reductive alkylation of ribonuclease has been reported.[78] In a subsequent study,[79] Jentoft and Dearborn characterized the inhibition by cyanide of reductive alkylation with sodium cyanoborohydride (Figure 43). This is of some importance since cyanide is a product of reductive alkylation with sodium cyanoborohydride. Inhibition by cyanide can be blocked by nickel(II) or cobalt(III). The observation that nickel(II) can preclude the inhibition of reductive alkylation by cyanide was shown to obviate the previously observed necessity for recrystallization of the sodium cyanoborohydride. The effect of NiCl$_2$ or KCN on the time course of the reductive methylation of bovine serum albumin at pH 7.5 (0.050 M HEPES) is shown in Figure 44. Additional studies on the development of reagents alternative to sodium borohydride have been reported from other laboratories. Feeney and co-workers[80] compared sodium cyanoborohydride, dimethylamine borane, and trimethylamine borane (Figure 45) with respect to effectiveness in reductive alkylation. Reduction at disulfide bonds was not observed with any of the three reagents. Dimethylamine borane was only slightly less effective than sodium cyanoborohydride while trimethylamine borane was much less effective (Figure 46). This decrease in effectiveness in reductive alkylation is balanced by the absence of toxic by-products such as cyanide evolving during the reaction. Figure 47 compares dimethylamine borane and trimethylamine borane in the reductive methylation of turkey ovomucoid in 0.2 M sodium phosphate, pH 7.0. Quantitative reductive methylation (equal to or greater than one methyl group per lysyl residue) is achieved at 10 mM formaldehyde with dimethylamine borane and at 50 mM formaldehyde with trimethylamine borane. It should be noted that a similar extent of modification is obtained with 5 mM formaldehyde using sodium cyanoborohydride. In a subsequent study[81] this laboratory reported the successful use of pyridine borane in the reductive alkylation of proteins. Wu and Means[82] have used reductive alkylation with a nonpolar aldehyde (dodecylaldehyde) to subsequently prepare insoluble proteins by binding of the modified protein to octyl-Sepharose.

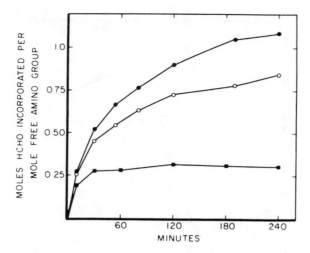

FIGURE 44. The effect of Ni(II) and KCN on the time course of the reductive methylation of albumin. The reactions were performed in 0.05 *M* HEPES, pH 7.5 with 10 m*M* sodium cyanoborohydride, and 2 m*M* [14C] formaldehyde and incubated in the presence of either 2 m*M* NiCl$_2$ (●) or 2 mM KCN (■) or no addition (○). (From Jentoft, N. and Dearborn, D. G., *Analyt. Biochem.*, 106, 186, 1980. With permission.)

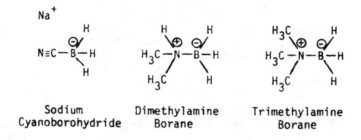

| Sodium Cyanoborohydride | Dimethylamine Borane | Trimethylamine Borane |

FIGURE 45. The structure of sodium cyanoborohydride, dimethylamine borane, and trimethylamine borane. (From Geoghegan, K. F., Cabacungan, J. C., Dixon, H. B. F., and Feeney, R. E., *Int. J. Peptide Protein Res.*, 17, 345, 1981. With permission.)

The reaction of glyceraldehyde with carbonmonoxyhemoglobin S has been explored by Acharya and Manning.[83] This reaction was performed with 0.010 *M* glyceraldehyde in phosphate-buffered saline, pH 7.4, and the resultant Schiff bases were stabilized by reduction with sodium borohydride (Figure 48). Using radiolabeled glyceraldehyde, these investigators were able to obtain support for the concept that there is selectivity in the reaction of sugar aldehydes with hemoglobin. The reaction product between glyceraldehyde and hemoglobin S did have stability properties without reduction that were not consistent with only Schiff base products. These investigators suggested that the glyceraldehyde-hemoglobin Schiff base could undergo an Amadori rearrangement (Figure 49) to form a stable ketoamine adduct which could be reduced with sodium borohydride to form a product identical to that obtained by direct reduction of the Schiff base. In a subsequent study, these investigators did show that the glyceraldehyde-hemoglobin S Schiff base could rearrange to form a ketamine via an Amadori rearrangement.[84] These investigators were able to use reaction with phenylhydrazine to detect the protein-bound ketamine adduct as shown in Figure 50.

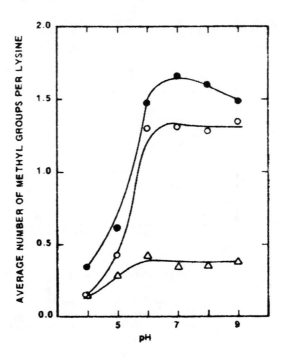

FIGURE 46.   The effect of pH on the reductive methylation of turkey ovomucoid in the presence of various reducing agents. The concentration of turkey ovomucoid was 5 mg/mℓ, the concentration of formaldehyde was 20 mM in the presence of either sodium cyanoborohydride (●), 15 mM, dimethylamine borane (○), 15 mM, or trimethylaminoborane (△), 15 mM. (From Geoghegan, K. F., Cabacungan, J. C., Dixon, H. B. F., and Feeney, R. E., *Int. J. Peptide Protein Res.*, 17, 345, 1981. With permission.)

Another class of aldehydes that react with protein to give interesting products are simple monosaccharides which exist in solution in enol and keto forms (Figure 51). Wilson[85] showed that bovine pancreatic ribonuclease dimer would react with lactose in the presence of sodium cyanoborohydride to yield an active derivative that shows selectivity in uptake by the liver during in vivo experiments. The modification of ribonuclease dimer was performed in 0.2 M potassium phosphate, pH 7.4 (phosphate buffer was used to protect lysine-41 from modification) at 37°C for 5 days with lactose and sodium cyanoborohydride. Under these conditions, 80% of the amino groups were modified. Bunn and Higgins[86] have explored the reaction of monosaccharides with protein amino groups in the presence of sodium cyanoborohydride in some detail. These investigators studied the reaction of hemoglobin with various monosaccharides in Krebs-Ringer phosphate buffer, pH 7.3 (Figure 52). The extent of modification was determined using tritiated sodium cyanoborohydride. The rate of modification was demonstrated to be a direct function of the amount of each sugar in the carbonyl (or keto) form (Figure 53). Thus the k_1 ($\times$ 10^{-3} mM^{-1} hr^{-1}) for D-glucose is 0.6 with 0.002% in the carbonyl form while the k_1 ($\times$ 10^{-3} mM^{-1} hr^{-1}) for D-ribose is 10.0 with 0.05% in the carbonyl form.

The reaction of trinitrobenzenesulfonic acid with amino groups has been of particular value in studying the function and reactivity of the ϵ-amino groups of lysyl residues in proteins.[87-90] The reaction of trinitrobenzenesulfonic acid with the primary amino groups in proteins is shown in Figure 54. The modification of amino groups with trinitrobenzenesul-

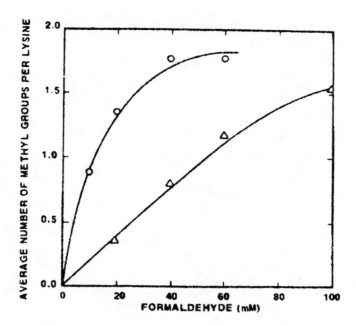

FIGURE 47. The effect of formaldehyde concentration on the reductive methylation of turkey ovomucoid. The reaction was performed in 0.2 M sodium phosphate, pH 7.0 with a turkey ovomucoid concentration of 5 mg/mℓ. The reducing agents were used at a concentration of 15 mM and included dimethylamine borane (○) and trimethylamine borane (△) at 22°C. The reducing agents were dissolved in methanol at a concentration of 150 mM and diluted 1:10 in the reaction mixture such that the final concentration of methanol was 10% (v/v). (From Geoghegan, K. F., Cabacungan, J. C., Dixon, H. B. F., and Feeney, R. E., *Int. J. Peptide Protein Res.*, 17, 345, 1981. With permission.)

fonic acid is easy to monitor by spectral analysis. In the presence of an excess of sulfite, absorbance at 420 nm is the most sensitive index, having ϵ = 2.0 × 10⁴ M^{-1} cm⁻¹. Absorbance at 420 nm is dependent upon the ability of the reaction product to form a complex with sulfite. It has proved convenient in our laboratories to use the fact that the spectrum of a trinitrobenzyl amino compound has an isosbestic point at 367 nm with ϵ = 1.05 × 10⁴ M^{-1} cm⁻¹. As suggested by Fields,[90] we recrystallize trinitrobenzenesulfonic acid from 2.0 M HCl prior to use. We generally perform the modifications in phosphate buffer (pH 6.0 to 9.0). The derivatives of α-amino groups and ε-amino groups have similar spectra with the exception that α-amino derivatives have a slightly higher extinction coefficient at 420 nm (ϵ = 2.20 × 10⁴ M^{-1} cm⁻¹) than ε-amino groups (ϵ = 1.92 × 10⁴ M^{-1} cm⁻¹).[99] Both of these derivatives have much higher extinction coefficients than the derivative obtained by reaction of trinitrobenzenesulfonic acid with cysteinyl residues (ϵ = 2.25 × 10³ M^{-1} cm⁻¹). The α-amino and ε-amino derivatives can be differentiated by their stability to acid or base hydrolysis. The α-amino derivatives are unstable to acid hydrolysis (8 hr at 110°C) or base hydrolysis.[91]

Frieden and co-workers have explored the reaction of trinitrobenzenesulfonic acid with bovine liver glutamate dehydrogenase.[92,93] In these studies, the modification was performed in 0.04 M potassium phosphate, pH 8.0 (Figure 55). Under these reaction conditions, the cysteinyl residues were not modified. The preparative reactions were terminated by reaction with β-mercaptoethanol. The reaction products were subjected to proteolytic digestion and certain of the sites of modification were identified as shown in Figure 56. It is of interest

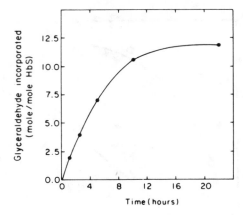

FIGURE 48.   Time course for the modification of the amino groups of carbonmonoxyhemoglobin S (the carbonmonoxy derivative of sickle cell hemoglobin). Carbonmonoxyhemoglobin (approximately 0.5 *M* with respect to tetramer) was combined with [^{14}C] glyceraldehyde (final concentration 10 m*M*) at pH 7.4 in phosphate-buffered saline at 37°C. Portions of the reaction mixture were removed at the indicated times, reduced with sodium borohydride and dialyzed vs. phosphate-buffered saline. The extent of modification was assessed by radiolabel incorporation. (From Acharya, A. S. and Manning, J. M., *J. Biol. Chem.*, 255, 1406, 1980. With permission.)

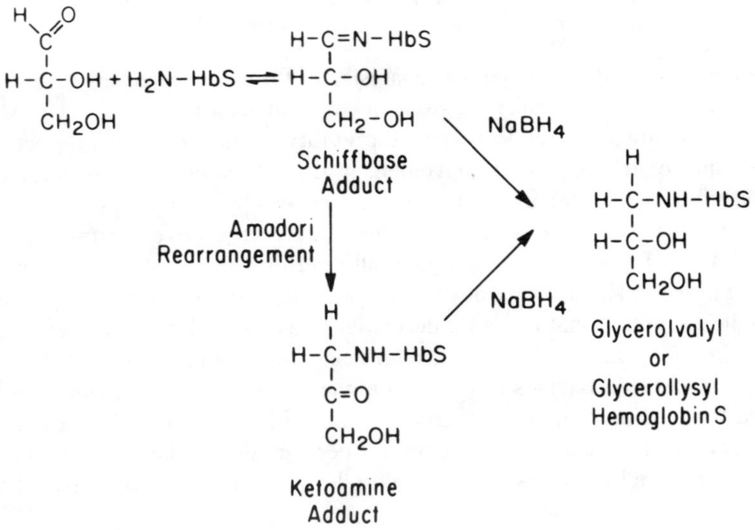

FIGURE 49.   Schematic representation of the formation of glycerovaline or glycerollysine on the reaction of hemoglobin S with glyceraldehyde. (From Acharya, A. S. and Manning, J. M., *J. Biol. Chem.*, 255, 1406, 1980. With permission.)

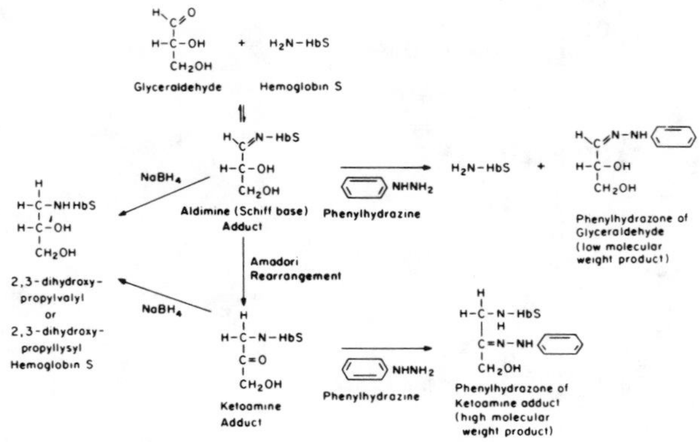

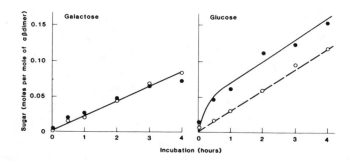

FIGURE 50. The reaction of the glyceraldehyde-hemoglobin adduct with either phenylhydrazine or sodium borohydride. (From Acharya, A. S. and Manning, J. M., *J. Biol. Chem.*, 255, 1406, 1980. With permission.)

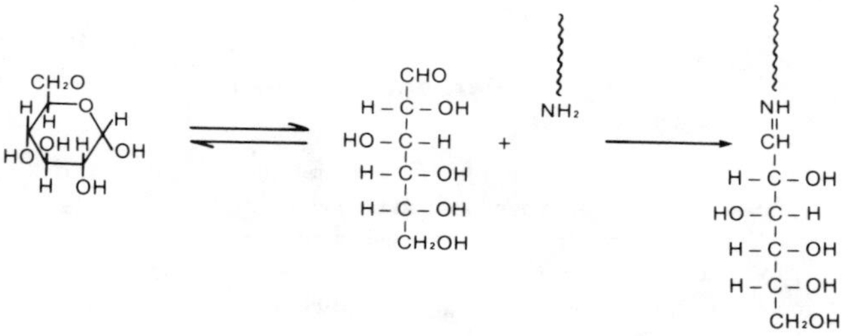

FIGURE 51. A scheme for the reaction of a monosaccharide with a primary amino group.

FIGURE 52. The measurement of the rate of condensation of monosaccharides with hemoglobin. The extent of reaction was measured either by incubation with unlabeled sugar followed by reduction of the aldimine linkage with tritiated sodium cyanoborohydride (open circles) or by incubation of the [^{14}C]-labeled sugar with hemoglobin followed by reduction with unlabeled sodium cyanoborohydride (closed circles). The left panel shows the rate of reaction with 42 mM D-galactose (k_1 = 1.9 × 10^{-3} mM^{-1} hr^{-1}. The right panel shows the reaction with 42 mM D-glucose (k_1 = 0.6 × 10^{-3} mM^{-1} hr^{-1}. The initial rapid rate of incorporation of D-[^{14}C] glucose can be explained by the small amount of rapidly reacting impurity remaining in the preparation. (From Bunn, H. F. and Higgins, R. J., *Science*, 213, 222, 1981. With permission.)

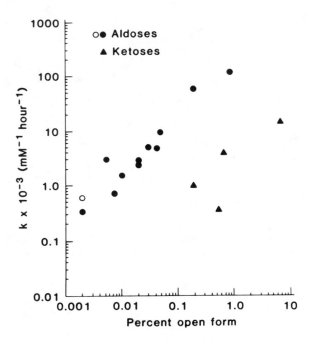

FIGURE 53.   The relation between the rate of condensation of monosaccharide with hemoglobin and the equilibrium between the open and ring structures of the monosaccharide (l_1). The open circle is for glucose ($k_1 = 0.6 \times 10^{-3}$ mM^{-1} hr^{-1}). The closed circles represent data for other aldoses and the closed triangles for ketoses. (From Bunn, H. F. and Higgins, P. J., *Science,* 213, 222, 1981. With permission.)

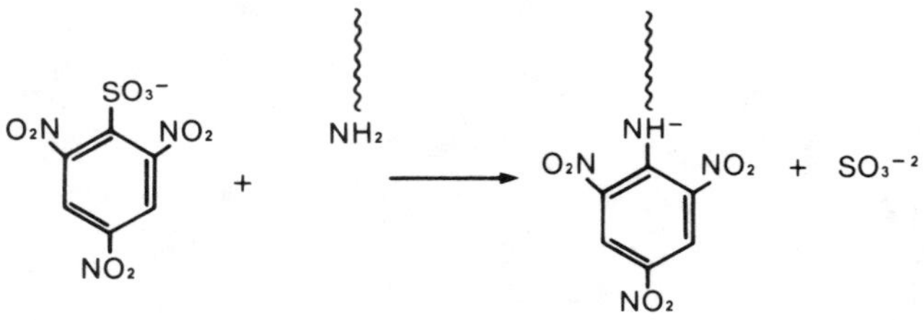

FIGURE 54.   The reaction of 2,4,6-trinitrobenzenesulfonic acid with primary amines in proteins.

that under certain conditions (with reduced coenzyme), glutamate dehydrogenase catalyzed the conversion of trinitrobenzenesulfonic acid to trinitrobenzene.[94]

The reaction of trinitrobenzenesulfonic acid with simple amines and hydroxide ions has been studied in some detail by Means and co-workers.[95] The reaction of trinitrobenzenesulfonic acid with hydroxide is first-order with respect to both trinitrobenzenesulfonate and hydroxide ions. Reaction with amines was considered in some detail. In general, reactivity of trinitrobenzenesulfonate with amines increases with increasing basicity except that secondary amines and *t*-alkylamines are comparatively unreactive. The specific binding of trinitrobenzenesulfonate to proteins must be considered in the study of the reaction of this compound with proteins. Only amines with a pKa greater than 8.7 follow a simple rate law. These investigators presented the following considerations regarding the reaction of trini-

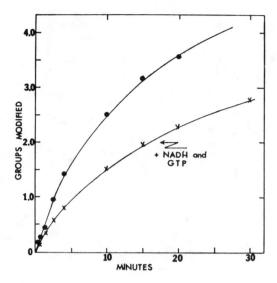

FIGURE 55. The time course of the reaction of 2,4,6-trinitrophenylsulfonic acid (TNBS) with bovine liver glutamate dehydrogenase. The reaction was performed in 0.04 M sodium phosphate, pH 8.0 at 25°C with a protein concentration of 1.0 mg/mℓ. The concentration of TNBS was 1 mM in the presence(x) and absence ($\bullet$) of NADH (600 μM) and GTP (500 μM). The extent of modification was determined by the increase in absorbance at 367 nm (= $1.05 \times 10^4\ M^{-1}$) in the absence of NADH and GTP and at 420 nm in the presence of NADH and GTP. This is required because of the strong absorbance of NADH at 367 nm. (From Coffee, C. J., Bradshaw, R. A., Goldin, B. R., and Frieden, C., *Biochemistry*, 10, 3516, 1971. With permission.)

trobenzenesulfonic acid with proteins. Reactivity is a sensitive measure of the basicity of an amino group. Adjacent charged groups have an influence on the rate of reaction with an increase observed with a positively charged group and a decrease with a negatively charged group. Proximity to surface hydrophobic regions which can bind trinitrobenzenesulfonic acid can increase the observed reactivity of a particular amino group.

Flügge and Heldt have explored the labeling of a specific membrane component with trinitrobenzenesulfonic acid[96] and pyridoxal-5'-phosphate.[97] The modification of the phosphate translocation protein in spinach chloroplasts with trinitrobenzenesulfonic acid was performed in 0.050 M HEPES, 0.33 M sorbitol, 0.001 MgCl$_2$, 0.001 M MnCl$_2$, 0.002 M EDTA, pH 7.6 at 4°C for periods of time up to 15 min at which point tritiated sodium borohydride was added to both terminate the reaction and radiolabel the trinitrophenyl derivatives.[98] It is possible to label components on the surface of membranes with trinitrobenzenesulfonic acid as the sulfonate moeity does not permit membrane penetration. The same is true for pyridoxal-5'-phosphate.

George and Borders[99] have examined the reaction of yeast enolase with trinitrobenzenesulfonic acid in 0.050 M Bicine (N,N-bis(2-hydroxyethyl) glycine), 0.001 M MgCl$_2$, 0.01 mM EDTA, pH 8.3. The extent of modification was determined spectrophotometrically at 367 nm ($\epsilon = 1.1 \times 10^4\ M^{-1}\ cm^{-1}$). Under these conditions, complete inactivation was obtained after the modification of 18 lysyl residues per subunit of yeast enolase.

On occasion, the modification of an amino acid residue in a protein is associated with an

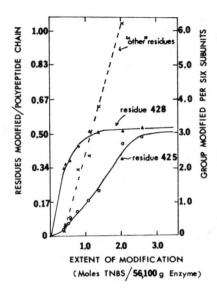

FIGURE 56.   The stoichiometry of the modification of lysine residues by 2,4,6-trinitrobenzenesulfonic acid in bovine glutamate dehydrogenase as a function of the total modification of the protein by TNBS. Note that residue 425 has been revised to residue 419 and residue 428 to residue 422 from more recent primary structure analysis. The assignment of the extent of modification was obtained by analysis of peptides derived from the modified protein, either trypsin, thermolysin or chymotrypsin. (From Coffee, C. J., Bradshaw, R. A., Goldin, B. R., and Frieden, C., *Biochemistry*, 10, 3516, 1971. With permission.)

apparent increase in catalytic activity. This was the situation with the modification of 14S and 30S dynein adenosine triphosphatase activities with trinitrobenzenesulfonic acid.[100] In this study, the reaction was performed in 0.030 M barbital, pH 8.5 at 25°C. The extent of modification was determined spectrophotometrically at 345 nm (ϵ = 1.45 × 10^4 M^{-1} cm^{-1}). In studies similar to those obtained with glutamate dehydrogenase as discussed above,[94] glutathione reductase was demonstrated to reduce trinitrobenzenesulfonate.[101] Inhibition of glutathione reductase was noted at low concentration (0.05 μM) of trinitrobenzenesulfonate. There was no significant effect on lipoamide dehydrogenase.

The reaction of myosin ATPase with trinitrobenzenesulfonate has been reported.[102] This modification was performed with a tenfold molar excess of trinitrobenzenesulfonate in 0.050 M Tris, pH 7.5 at 25°C either with or without 0.010 M adenosine triphosphate or 0.010 M magnesium adenosine triphosphate. It was possible to localize the site of modification to a specific domain.

The reaction of trinitrobenzene sulfonic acid with ammonium has also been investigated by Whitaker and co-workers.[103] This reaction was performed in tetraborate buffer and 1 μM sulfite. The rate of the reaction was determined by following the increase in absorbance at 420 nm (ϵ = 2.02 × 10^4 M^{-1} cm^{-1}). The rate of reaction with ammonium (k = 0.128 min^{-1}) was slower than that with the average amine in a protein (k = 0.907 min^{-1} for enterotoxin) (Figure 57). The reaction with ammonium does, however, provide a sensitive assay for ammonia (as low as 6 nmol) with a precision of 1 to 2%.

The use of trinitrobenzenesulfonate in the selective modification of membrane surface components has been explored by Salem and co-workers.[104] This study involved the modification of intact cells with the trinitrobenzenesulfonic acid (dissolved in methyl alcohol) diluted to a 1% methanolic solution. As mentioned above, trinitrobenzenesulfonate does not

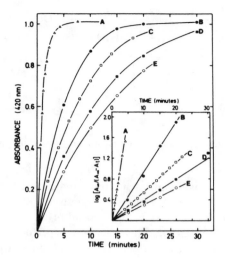

FIGURE 57. The reaction of 2,4,6-trinitrobenzenesulfonic acid (TNBS) with ammonia and the available amino groups of enterotoxin. The basic experimental approach was that described by Fields[90]. The rate assay approach involves the addition of TNBS to a reference cuvette containing only reagent and cuvette containing the sample. The difference in the rate of increase in absorbance at 420 nm is recorded (the buffer contains sodium sulfite). In the endpoint method, the reaction is allowed to proceed for a period of time at which point it is terminated by the addition of 0.1 M NaH$_2$PO$_4$ containing 1.5 mM sodium sulfite. The absorbance at 420 nm is then determined. Experiment A used the rate assay method with 6.7 μM enterotoxin at 25°C. Curve B used the endpoint method with 50 μM ammonium sulfate at 35°C. Curve C used the rate assay method with 50 μM ammonium sulfate at 25°C. Curve D used the endpoint method with 50 μM ammonium sulfate at 25°C with twice the TNBS concentration used in the other experiments. Curve E used the endpoint method with 50 μM ammonium sulfate. The insert is a plot of the rate data according to the first-order rate equation. (From Whitaker, J. R., Granum, P. E., and Ausen, G., *Analyt. Biochem.*, 108, 72, 1980. With permission.)

pass across (or into) membranes, being more hydrophilic than, for example, fluorodinitrobenzene.

Guanidation of proteins is a reaction which is fairly specific for ε-amino groups.[105] This modification involves the reaction of O-methylisourea with lysyl residues at basic pH (pH > 9) to yield homoarginine. This reaction is fairly slow and generally takes several days to go to completion. Awad and co-workers[106] have observed that the reaction of O-methylisourea with proteins at pH 10.5 results in stabilization. In this study, the modification was allowed to proceed for 4 days at 4°C. Bregman and co-workers[107] examined the modification of the single lysyl residue in glucagon. In this study 3.4 gm O-methylisourea (O-methylisourea hydrogen sulfate) was dissolved in 20 mℓ H$_2$O and 6.0 g Ba(OH)$_2$ added followed by filtration or centrifugation to remove the resulting BaSO$_4$. The pH of the solution was adjusted to 11.0 and 200 mg glucagon added. The reaction was allowed to proceed for 8 hr at 4°C and was terminated by the addition of glacial acetic acid. The products of the reaction were purified on Sulfopropyl-Sephadex.

The modification of lysyl residues with 2,4-pentanedione has been reported by Dunlap and co-workers.[108] This reaction is stated to be specific for lysyl residues. The reaction with 2,4-pentanedione was performed with 0.2 M reagent (a fresh 1.0 M solution in phosphate buffer is prepared daily) in 0.1 M potassium phosphate, 0.1 M KCl, 20% glycerol, pH 7.0. The extent of modification is determined by spectrophotometric analysis of the enamine product of the reaction at 312 nm (ε = 2.0 × 10^4 M^{-1} cm^{-1}). Protein concentration can be determined by absorbance at 278 nm after correction for absorbance by the enamine derivatives at that wavelength. The modified enzyme could be reactivated by incubation in phosphate buffer, pH 7.0 at 30°C but not at 4°C. Neutral hydroxylamine at 25, 33, or 40°C did not result in reactivation but did cause loss of absorbance at 312 nm.

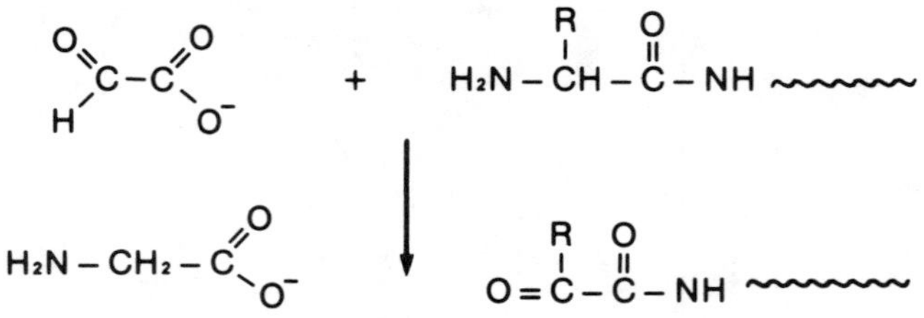

FIGURE 58. The modification of α-amino groups in proteins with glyoxylate.

The modification of proteins by enzymatic transamination has been reported.[109]

It is possible to selectively modify the α-amino groups of proteins by chemical transamination with glyoxylate (Figure 58) at slightly acid pH.[110-111] This modification has been applied to *Euglena* cytochrome C-552. This reaction was performed in 2.0 *M* sodium acetate, 0.10 *M* acetic acid, 0.005 *M* nickel sulfate, 0.2 *M* sodium glyoxylate and resulted in the complete loss of the amino terminal residue. Snake venom phospholipase A$_2$ has been subjected to chemical transamination.[111] This reaction was performed in 2.0 *M* sodium acetate, 0.4 *M* acetic acid, 0.010 *M* cupric ions, 0.1 *M* glyoxylic acid, pH 5.5.

REFERENCES

1. **Gurd, F. R. N.**, Carboxymethylation, *Meth. Enzymol.*, 11, 532, 1967.
2. **Sanger, F. and Tuppy, H.**, The amino acid sequence in the phenylalanyl chain of insulin. I. The identification of lower peptides from partial hydrolysates, *Biochem. J.*, 49, 463, 1951.
3. **Carty, R. P. and Hirs, C. H. W.**, Modification of bovine pancreatic ribonuclease A with 4-sulfonyloxy-2-nitrofluorobenzene, *J. Biol. Chem.*, 243, 5254, 1968.
4. **Franklin, J. G. and Leslie, J.**, Some enzymatic properties of trypsin after reaction with 1-dimethylaminonaphthalene-5-sulfonyl chloride, *Can. J. Biochem.*, 49, 516, 1971.
5. **Bello, J., Iijima, H., and Kartha, G.**, A new arylating agent, 2-carboxy-4,6-dinitrochlorobenzene. Reaction with model compounds and bovine pancreatic ribonuclease, *Int. J. Peptide Protein Res.*, 14, 199, 1979.
6. **Hiratsuka, T. and Uchida, K.**, Lysyl residues of cardiac myosin accessible to labeling with a fluorescent reagent, *N*-methyl-2-anilino-6-naphthalenesulfonyl chloride, *J. Biochem.*, 88, 1437, 1980.
7. **Cory, R. P., Becker, R. R., Rosenbluth, R., and Isenberg, I.**, Synthesis and fluorescent properties of some *N*-methyl-2-anilino-6-naphthalensulfonyl derivatives, *J. Am. Chem. Soc.*, 90, 1643, 1968.
8. **Turner, D. C. and Brand, L.**, Quantitative estimation of protein binding site polarity. Fluorescence of *N*-arylaminonaphthalenesulfonates, *Biochemistry*, 7, 3381, 1968.
9. **Welches, W. R. and Baldwin, T. O.**, Active center studies on bacterial luciferase: modification of the enzyme with 2,4-dinitrofluorobenzene, *Biochemistry*, 20, 512, 1981.
10. **Riordan, J. F. and Vallee, B. L.**, Acetylation, *Meth. Enzymol.*, 11, 565, 1967.
11. **Merle, M., Lefevre, G., Staub, J. F., Raulais, D., and Milhaud, G.**, Acylation of porcine and bovine calcitonin: effects on hypocalcemic activity in the rat, *Biochem. Biophys. Res. Commun.*, 79, 1071, 1977.
12. **Aviram, I.**, The role of lysines in *Euglena* cytochrome C-552. Chemical modification studies, *Arch. Biochem. Biophys.*, 181, 199, 1977.
13. **Karibian, D., Jones, C., Gertler, A., Dorrington, K. J., and Hofmann, T.**, On the reaction of acetic and maleic anhydrides with elastase. Evidence for a role of the NH$_2$-terminal valine, *Biochemistry*, 13, 2891, 1974.
14. **Smith, M. B., Stonehuerner, J., Ahmed, A. J., Staudenmayer, N., and Millett, F.**, Use of specific trifluoroacetylation of lysine residues in cytochrome C to study the reaction with cytochrome b$_5$, cytochrome c$_1$ and cytochrome oxidase, *Biochem. Biophys. Acta*, 592, 303, 1980.
15. **Webb, M., Stonehuerner, J., and Millett, F.**, The use of specific lysine modifications to locate the reaction site of cytochrome C with sulfite oxidase, *Biochim. Biophys. Acta*, 593, 290, 1980.

16. **Ahmed, A. J. and Millett, F.,** Use of specific lysine modifications to identify the site of reaction between cytochrome C and ferricyanide, *J. Biol. Chem.,* 256, 1611, 1981.
17. **Smith, M. B. and Millett, F.,** A ^{19}F nuclear magnetic resonance study of the interaction between cytochrome C and cytochrome C peroxidase, *Biochem. Biophys. Acta,* 626, 64, 1980.
18. **Hitchcock, S. E., Zimmerman, C. J., and Smalley, C.,** Study of the structure of troponin-T by measuring the relative reactivities of lysines with acetic anhydride, *J. Mol. Biol.,* 147, 125, 1981.
19. **Hitchcock, S. E.,** Study of the structure of troponin-C by measuring the relative reactivities of lysines with acetic anhydride, *J. Mol. Biol.,* 147, 153, 1981.
20. **Klotz, I. M.,** Succinylation, *Meth. Enzymol.,* 11, 576, 1967.
21. **Meighen, E. A., Nicolim, M. Z., and Hustings, J. W.,** Hybridization of bacterial luciferase with a variant produced by chemical modification, *Biochemistry,* 10, 4062, 1971.
22. **Shetty, K. J. and Rao, M. S. N.,** Effect of succinylation on the oligomeric structure of arachin, *Int. J. Peptide Protein Res.,* 11, 305, 1978.
23. **Shiao, D. D. F., Lumry, R., and Rajender, S.,** Modification of protein properties by change in charge. Succinylated chymotrypsinogen, *Eur. J. Biochem.,* 29, 377, 1972.
24. **Atassi, M. Z. and Habeeb, A. F. S. A.,** Reaction of protein with citraconic anhydride, *Meth. Enzymol.,* 25, 546, 1972.
25. **Habeeb, A. F. S. A. and Atassi, M. Z.,** Enzymic and immunochemical properties of lysozyme. Evaluation of several amino group reversible blocking reagents, *Biochemistry,* 9, 4939, 1970.
26. **Shetty, J. K. and Kinsella, J. E.,** Ready separation of proteins from nucleoprotein complexes by reversible modification of lysine residues, *Biochem. J.,* 191, 269, 1980.
27. **Weisgraber, K. H., Innerarity, T. L., and Mahley, R. W.,** Role of the lysine residues of plasma lipoproteins in high affinity binding to cell surface receptors on human fibroblasts, *J. Biol. Chem.,* 253, 9053, 1978.
28. **Mahley, R. W., Weisgraber, K. H., Innerarity, T. L., and Windmueller, H. G.,** Accelerated clearance of low-density and high-density lipoproteins and retarded clearance of E apoprotein-containing lipoproteins from the plasma of rats after modification of lysine residues, *Proc. Natl. Acad. Sci. U.S.A.,* 76, 1746, 1979.
29. **Urabe, I., Yamamoto, M., Yamada, Y., and Okada, H.,** Effect of hydrophobicity of acyl groups on the activity and stability of acylated thermolysin, *Biochim. Biophys. Acta,* 524, 435, 1978.
30. **Howlett, G. J. and Wardrop, A. J.,** Dissociation and reconstitution of human erythrocyte membrane proteins using 3,4,5,6-tetrahydrophthalic anhydride, *Arch. Biochem. Biophys.,* 188, 429, 1978.
31. **Stark, G. R., Stein, W. H., and Moore, S.,** Reaction of the cyanate present in aqueous urea with amino acids and proteins, *J. Biol. Chem.,* 235, 3177, 1960.
32. **Stark, G. R. and Smyth, D. G.,** The use of cyanate for the determination of NH_2-terminal residues in proteins, *J. Biol. Chem.,* 238, 214, 1963.
33. **Stark, G. R.,** Modification of proteins with cyanate, *Meth. Enzymol.,* 25, 579, 1972.
34. **Shaw, D. C., Stein, W. H., and Moore, S.,** Inactivation of chymotrypsin by cyanate, *J. Biol. Chem.,* 239, PC 671, 1964.
35. **Cerami, A. and Manning, J. M.,** Potassium cyanate as an inhibitor of the sickling of erythrocytes *in vitro, Proc. Natl. Acad. Sci. U.S.A.,* 68, 1180, 1971.
36. **Lee, C. K. and Manning, J. M.,** Kinetics of the carbamylation of the amino groups of sickle cell hemoglobin by cyanate, *J. Biol. Chem.,* 248, 5861, 1973.
37. **Njikam, N., Jones, W. M., Nigen, A. M., Gillette, P. N., Williams. R. C., Jr., and Manning, J. M.,** Carbamylation of the chains of hemoglobin S by cyanate *in vitro* and *in vivo, J. Biol. Chem.,* 248, 8052, 1973.
38. **Nigen, A. M., Bass, B. D., and Manning, J. M.,** Reactivity of cyanate with valine-1 (α) of hemoglobin. A probe of conformational change and anion binding, *J. Biol. Chem.,* 251, 7638, 1976.
39. **Plapp, B. V., Moore, S., and Stein, W. H.,** Activity of bovine pancreatic deoxyribonuclease A with modified amino groups, *J. Biol. Chem.,* 246, 939, 1971.
40. **Chollet, R. and Anderson, L. L.,** Cyanate modification of essential lysyl residues in the catalytic subunit of tobacco ribulosebisphosphate carboxylase, *Biochem. Biophys. Acta,* 525, 455, 1978.
41. **Kang, E. P.,** Amidinated thrombin: preparation and peptidase activity, *Thromb. Res.,* 12, 177, 1977.
42. **Sekiguchi, T., Oshiro, S., Goingo, E. M., and Nosoh, Y.,** Chemical modification of ϵ-amino groups in glutamine synthetase from *Bacillus stearothermophilus* with ethyl acetimidate, *J. Biochem.,* 85, 75, 1979.
43. **Browne, D. J. and Kent, S. B. H.,** Formation of non-amidine products in the reaction of primary amines with imido esters, *Biochem. Biophys. Res. Commun.,* 67, 126, 1975.
44. **Monneron, A. and d'Alayer, J.,** Effects of imido-esters on membrane-bound adenylate cyclase, *FEBS Lett.,* 122, 241, 1980.
45. **Plapp, B. V.,** Enhancement of the activity of horse liver alcohol dehydrogenase by modification of amino groups at the active sites, *J. Biol. Chem.,* 245, 1727, 1970.

46. **Dunathan, H. C.,** Stereochemical aspects of pyridoxal phosphate catalysis, in *Adv. Enzymol.,* 35, 79, 1971.
47. **Shapiro, S., Enser, M., Pugh, E., and Horecker, B. L.,** The effect of pyridoxal phosphate on rabbit muscle aldolase, *Arch. Biochem. Biophys.,* 128, 554, 1968.
48. **Schnackerz, K. D. and Noltmann, E. A.,** Pyridoxal-5′-phosphate as a site-specific protein reagent for a catalytically critical lysine residue in rabbit muscle phosphoglucose isomerase, *Biochemistry,* 10, 4837, 1971.
49. **Havran, R. T. and du Vigneaud, V.,** The structure of acetone-lysine vasopressin as established through its synthesis from the acetone derivative of S-benzyl-L-cysteinyl-L-tyrosine, *J. Am. Chem. Soc.,* 91, 2696, 1969.
50. **Paech, C. and Tolbert, N. E.,** Active site studies of ribulose-1,5-bisphosphate carboxylase/oxygenase with pyridoxal-5′-phosphate, *J. Biol. Chem.,* 253, 7864, 1978.
51. **Kent, A. B., Krebs, E. G., and Fischer, E. H.,** Properties of crystalline phosphorylase b, *J. Biol Chem.,* 232, 549, 1958.
52. **Wimmer, M. J., Mo, T., Sawyers, D. L., and Harrison, J. H.,** Biphasic inactivation of porcine heart mitochondrial malate dehydrogenase by pyridoxal-5′-phosphate, *J. Biol. Chem.,* 250, 710, 1975.
53. **Bleile, D. M., Jameson, J. L., and Harrison, J. H.,** Inactivation of porcine heart cytoplasmic malate dehydrogenase by pyridoxal-5′-phosphate, *J. Biol. Chem.,* 251, 6304, 1976.
54. **Jones, C. W., III and Priest, D. G.,** Interaction of pyridoxal-5′-phosphate with apo-serine hydroxymethyltransferase, *Biochim. Biophys. Acta,* 526, 369, 1978.
55. **Cortijo, M., Jimenez, J. S., and Lior, J.,** Criteria to recognize the structure and micropolarity of pyridoxal-5′-phosphate binding sites in proteins, *Biochem. J.,* 171, 497, 1978.
56. **Cake, M. A., DiSorbo, D. M., and Litwack, G.,** Effect of pyridoxal phosphate on the DNA binding site of activated hepatic glucocorticoid receptor, *J. Biol. Chem.,* 253, 4886, 1978.
57. **Slebe, J. C. and Martinez-Carrion, M.,** Selective chemical modification and ¹⁹F NMR in the assignment of a pK value to the active site lysyl residue in aspartate transaminase, *J. Biol Chem.,* 253, 2093, 1978.
58. **Nishigori, H. and Toft, D.,** Chemical modification of the avian progesterone receptor by pyridoxal-5′-phosphate, *J. Biol. Chem.,* 254, 9155, 1979.
59. **Sugiyama, Y. and Mukohata, Y.,** Modification of one lysine by pyridoxal phosphate completely inactivates chloroplast coupling factor 1 ATPase, *FEBS Lett.,* 98, 276, 1979.
60. **Peters, H., Risi, S., and Dose, K.,** Evidence for essential primary amino groups in a bacterial coupling factor F₁ ATPase, *Biochem. Biophys. Res. Commun.,* 97, 1215, 1980.
61. **Gould, K. G. and Engel, P. C.,** Modification of mouse testicular lactate dehydrogenase by pyridoxal 5′-phosphate, *Biochem. J.,* 191, 365, 1980.
62. **Ogawa, H. and Fujioka, M.,** The reaction of pyridoxal-5′-phosphate with an essential lysine residue of saccharopine dehydrogenase (L-lysine-forming), *J. Biol. Chem.,* 255, 7420, 1980.
63. **Forrey, A. W., Olsgaard, R. B., Nolan, C., and Fischer, E. H.,** Synthesis and properties of α- and ε-pyridoxyl lysines and their phosphorylated derivatives, *Biochemie,* 53, 269, 1971.
64. **Sober, H. A.,** in *Handbook of Biochemistry,* 2nd ed., The Chemical Rubber Company, Cleveland, Ohio, 1970.
65. **Moldoon, T. G. and Cidlowski, J. A.,** Specific modifications of rat uterine estrogen receptor by pyridoxal-5′-phosphate, *J. Biol. Chem.,* 2, 55, 3100, 1980.
66. **Ohsawa, H., and Gualerzi, C.,** Structure-function relationship in *Escherichia coli* inhibition factors. Identification of a lysine residue in the ribosomal binding site of initiation factor by site-specific chemical modification with pyrodixal phosphate, *J. Biol. Chem.,* 256, 4905, 1981.
67. **Bürger, E. and Görisch, H.,** Evidence for an essential lysine at the active site of L-histidinol:NAD⁺ oxidoreductase, a bifunctional dehydrogenase, *Eur. J. Biochem.,* 118, 125, 1981.
68. **Feeney, R. E., Blankenhorn, G., and Dixon, H. B. F.,** Carbonyl-amine reactions in protein chemistry, in *Adv. Protein Chem.,* 29, 135, 1975.
69. **Means, G. E.,** Reductive alkylation of amino groups, *Meth. Enzymol.,* 47, 469, 1977.
70. **Means, G. E. and Feeney, R. E.,** Reductive alkylation of amino groups in proteins, *Biochemistry,* 7, 2192, 1968.
71. **Rice, R. H., Means, G. E., and Brown, W. D.,** Stabilization of bovine trypsin by reductive methylation, *Biochim. Biophys. Acta,* 492, 316, 1977.
72. **Morris, R. W., Cagan, R. H., Martenson, R. E., and Deibler, G.,** Methylation of the lysine residues of monellin, *Proc. Soc. Exp. Biol. Med.,* 157, 194, 1978.
73. **Chen, F. M. F. and Benoiton, N. L.,** Reductive *N,N*-dimethylation of amino acid and peptide derivatives using methanol as the carbonyl source, *Can. J. Biochem.,* 56, 150, 1978.
74. **Dottavio-Martin, D. and Ravel, J. M.,** Radiolabeling of proteins by reductive alkylation with [¹⁴C] formaldehyde and sodium cyanoborohydride, *Analyt. Biochem.,* 87, 562, 1978.
75. **Fretheim, K., Iwai, S., and Feeney, R. F.,** Extensive modification of protein amino groups by reductive addition of different sized substitutents, *Int. J. Peptide Protein Res.,* 14, 451, 1979.

76. **Geoghegan, K. F., Ybarra, D. M., and Feeney, R. E.,** Reversible reductive alkylation of amino groups in proteins, *Biochemistry,* 18, 5392, 1979.

77. **Jentoft, N. and Dearborn, D. G.,** Labeling of proteins by reductive methylation using sodium cyanoborohydride, *J. Biol. Chem.,* 254, 4359, 1979.

78. **Jentoft, J. E., Jentoft, N., Gerken, T. A., and Dearborn, D. G.,** ^{13}C NMR studies of ribonuclease A methylated with [^{13}C] formaldehyde, *J. Biol. Chem.,* 254, 4366, 1979.

79. **Jentoft, N. and Dearborn, D. G.,** Protein labeling by reductive methylation with sodium cyanoborohydride: effect of cyanide and metal ions on the reaction, *Analyt. Biochem.,* 106, 186, 1980.

80. **Geoghegan, K. F., Cabacungan, J. C., Dixon, H. B. F., and Feeney, R. E.,** Alternative reducing agents for reductive methylation of amino groups in proteins, *Int. J. Peptide Protein Res.,* 17, 345, 1981.

81. **Cabacungan, J. C., Ahmed, A. J., and Feeney, R. E.,** Amine boranes as alternative reducing agents for reductive alkylation of proteins, *Anal. Biochem.,* 124, 272, 1982.

82. **Wu, H.-L. and Means, G. E.,** Immobilization of proteins by reductive alkylation with hydrophobic aldehydes, *Biotechnol. Bioeng.,* 23, 855, 1981.

83. **Acharya, A. S. and Manning, J. M.,** Reactivity of the amino groups of carbonmonoxyhemoglobin S with glyceraldehyde, *J. Biol. Chem.,* 255, 1406, 1980.

84. **Acharya, A. S., and Manning, J. M.,** Amadori rearrangement of glyceraldehyde-hemoglobin Schiff base adducts. A new procedure for the determination of ketoamine adducts in proteins, *J. Biol. Chem.,* 255, 7218, 1980.

85. **Wilson, G.,** Effect of reductive lactosamination on the hepatic uptake of bovine pancreatic ribunoclease A dimer, *J. Biol. Chem.,* 253, 2070, 1978.

86. **Bunn, H. F. and Higgins, P. J.,** Reaction of monosaccharides with proteins: possible evolutionary significance, *Science,* 213, 222, 1981.

87. **Goldfarb, A. R.,** A kinetic study of the reactions of amino acids and peptides with trinitrobenzenesulfonic acid, *Biochemistry,* 5, 2570, 1966.

88. **Goldfarb, A. R.,** Heterogeneity of amino groups in proteins. I. Human serum albumin, *Biochemistry,* 5, 2574, 1966.

89. **Habeeb, A. F. S. A.,** Determination of free amino groups in proteins by trinitrobenzenesulfonic acid, *Anal. Biochem.,* 14, 328, 1966.

90. **Fields, R.,** The rapid determination of amino groups with TNBS, *Meth. Enzymol.,* 25, 464, 1972.

91. **Kotaki, A. and Satake, K.,** Acid and alkaline degradation of the TNP-amino acids and peptides, *J. Biochem.,* 56, 299, 1964.

92. **Coffee, C. J., Bradshaw, R. A., Goldin, B. R., and Frieden, C.,** Identification of the sites of modification of bovine liver glutamate dehydrogenase reacted with trinitrobenzenesulfonate, *Biochemistry,* 10, 3516, 1971.

93. **Goldin, B. R. and Frieden, C.,** Effects of trinitrophenylation of specific lysyl residues on the catalytic, regulatory and molecular properties of bovine liver glutamate dehydrogenase, *Biochemistry,* 10, 3527, 1971.

94. **Bates, D. J., Goldin, B. R., and Frieden, C.,** A new reaction of glutamate dehydrogenase: the enzyme-catalyzed formation of trinitrobenzene from TNBS in the presence of reduced coenzyme, *Biochem. Biophys. Res. Commun.,* 39, 502, 1970.

95. **Means, G. E., Congdon, W. I., and Bender, M. L.,** Reactions of 2,4,6-trinitrobenzenesulfonate ion with amines and hydroxide ion, *Biochemistry,* 11, 3564, 1972.

96. **Flügge, U. I. and Heldt, H. W.,** Specific labelling of the active site of the phosphate translocator in spinach chloroplasts by 2,4,6-trinitrobenzene sulfonate, *Biochem. Biophys. Res. Commun.,* 84, 37, 1978.

97. **Flügge, U. I. and Heldt, H. W.,** Specific labelling of a protein involved in phosphate transport of chloroplasts by pyridoxal-5′-phosphate, *FEBS Lett.,* 82, 29, 1977.

98. **Parrott, C. L. and Shifrin, S.,** A spectrophotometric study of the reaction of borohydride with trinitrophenyl derivatives of amino acids and proteins, *Biochim. Biophys. Acta,* 491, 114, 1977.

99. **George, A. L., Jr. and Borders, C. L., Jr.,** Chemical modification of histidyl and lysyl residues in yeast enolase, *Biochim. Biophys. Acta,* 569, 63, 1979.

100. **Shimizu, T.,** Enhancement of 14S and 30S dynein adenosine triphosphatase activities by modification of amino groups with trinitrobenzenesulfonate. A comparison with modification of SH groups, *J. Biochem.,* 85, 1421, 1979.

101. **Carlberg, I. and Mannervik, B.,** Interaction of 2,4,6-trinitrobenzene sulfonate and 4-chloro-7-nitrobenzo-2-oxa-1,3-diazole with the active sites of glutathione reductase and lipoamide dehydrogenase, *Acta Chem. Scand.,* B34, 144, 1980.

102. **Mornet, D., Pantel, P., Bertrand, R., Audemard, E., and Kassab, R.,** Localization of the reactive trinitrophenylated lysyl residue of myosin ATPase site in the NH_2-terminal (27K domain) of S1 heavy chain, *FEBS Lett.,* 117, 183, 1980.

103. **Whitaker, J. R., Granum, P. E., and Aasen, G.,** Reaction of ammonia with trinitrobenzene sulfonic acid, *Analyt. Biochem.,* 108, 72, 1980.

104. **Salem, N., Jr., Lauter, C. J., and Trams, E. G.,** Selective chemical modification of plasma membrane ectoenzymes, *Biochim. Biophys. Acta,* 641, 366, 1981.
105. **Kimmel, J. R.,** Guanidination of proteins, *Meth. Enzymol.,* 11, 584, 1967.
106. **Cupo, P., El-Deiry, W., Whitney, P. L., and Awad, W. M., Jr.,** Stabilization of proteins by guanidation, *J. Biol. Chem.,* 255, 10828, 1980.
107. **Bregman, M. D. Trivedi, D., and Hruby, V. J.,** Glucagon amino groups. Evaluation of modification leading to antagonism and agonism, *J. Biol. Chem.,* 255, 11725, 1980.
108. **Otwell, H. B., Cipollo, K. L., and Dunlap, R. B.,** Modification of lysyl residues of dihydrofolate reductase with 2,4-pentanedione, *Biochim. Biophys. Acta,* 568, 297, 1979.
109. **Iwanij, V.,** The use of liver transglutaminase for protein labeling, *Eur. J. Biochem.,* 80, 359, 1977.
110. **Dixon, H. B. F. and Fields, R.,** Specific modification of NH$_2$-terminal residues by transamination, *Meth. Enzymol.,* 25, 409, 1972.
111. **Verheij, H. M., Egmond, M. R., and de Haas, G. H.,** Chemical modification of the α-amino group in snake venom phospholipases A$_2$. A comparison of the interaction of pancreatic and venom phospholipases with lipid-water interfaces, *Biochemistry,* 20, 94, 1981.

INDEX

A

B